Essential Ornithology

Essential Ornithology

Graham Scott

Department of Biological Sciences, University of Hull, UK

OXFORD
UNIVERSITY PRESS

OXFORD
UNIVERSITY PRESS

Great Clarendon Street, Oxford ox2 6DP

Oxford University Press is a department of the University of Oxford.
It furthers the University's objective of excellence in research, scholarship,
and education by publishing worldwide in

Oxford   New York

Auckland Cape Town Dar es Salaam Hong Kong Karachi
Kuala Lumpur Madrid Melbourne Mexico City Nairobi
New Delhi Shanghai Taipei Toronto

With offices in

Argentina Austria Brazil Chile Czech Republic France Greece
Guatemala Hungary Italy Japan Poland Portugal Singapore
South Korea Switzerland Thailand Turkey Ukraine Vietnam

Oxford is a registered trade mark of Oxford University Press
in the UK and in certain other countries

Published in the United States
by Oxford University Press Inc., New York

British Library Cataloguing in Publication Data
Data available

Library of Congress Cataloging-in-Publication Data
Scott, Graham (Graham W.)
Essential ornithology/Graham Scott.
p. cm.
ISBN 978-0-19-856997-8 (Pbk.)
1. Ornithology. 2. Birds. I. Title.
QL673.S275 2010
598—dc22   2010010914

Typeset by SPI Publisher Services, Pondicherry, India
Printed in Great Britain by Clays Ltd, St Ives plc

ISBN 978-0-19-856998-5 (Hbk.)
      978-0-19-856997-8 (Pbk.)

9 10 8

*For Lisa, William, and Adam who have been very, very patient,
and for Les and Shirley who help them to keep me birding!*

*In memory of my father William Scott. He encouraged my interest in
birds and for that I shall be eternally grateful.*

Contents

Preface

This book is intended to be an introduction to what I consider to be the essentials of ornithology; those areas of the broader biology of birds that are, in my view, the minimum that a student of ornithology should know. In writing the text I am aware that in places I could have included more depth, and that in places I could have provided a broader treatment, but my aim was to produce a relatively short useful text and so I was unable to follow through every avenue of thought—no matter how much I should have liked to!

Instead, by covering a breadth of material and by restricting myself to case studies that should be accessible to all, I hope that I have provided the reader with a 'way in' to academic ornithology.

Acknowledgements

A book such as this one may have a single author—but it would be impossible to write without access to all of the published works produced by active researchers/writers in ornithology; I am indebted to them all. Similarly I owe a debt of thanks to friends and colleagues who have provided their views on sections of the text, provided most of the pictures that are included in it, and assisted in the publication progress. At the risk of missing anyone of them, they are: Lisa Scott, Phil Wheeler, James Spencer, Robin Arundale, Peter Dunn, Ian Robinson, Ian Grier, William Scott, Les Hatton, Chris Redfern, Shirley Millar, Andy Gosler, Sjirk Geets, Stanislav Pribil, Ian Sherman, and Helen Eaton.

CHAPTER 1

Evolution of birds

Creation is never over. It had a beginning but it has no ending.

Immanuel Kant, A General Natural
History of the Heavens, 1755

I often ask my students, 'what is a bird?', it's a question that I was asked when I was a student and then, as now, the most common answer is 'a flying vertebrate'. Well of course this is only partly right. Birds, like fish, amphibians, reptiles, and mammals, do have vertebrae (and the other defining characteristics of that group) and are therefore vertebrates—but not all birds fly, nor are all flying vertebrates birds (what about bats?). In practice it can be surprisingly difficult to define a bird, although we all know what one is. A feathered vertebrate may do the job because the only extant vertebrates to have feathers are the birds. Feathers, however, can no longer be claimed to belong uniquely to the birds following the discovery of fossilized specimens of a number of feathered dinosaurs such as *Microraptor* and *Sinosauropteryx*. Many of the other defining features of birds are in fact shared with one or more vertebrate groups and so only set birds apart when considered in concert. Like most reptiles they lay eggs and their young develop outside of the body of the parent. Like crocodilians and mammals they have a four-chambered heart. Like mammals they are warm blooded and have a high metabolic rate which in turn requires that they feed regularly. Uniquely they lack teeth and have a beak. They have bones that are pneumatized (they have air-filled cavities) and are therefore light compared with the solid bones of other vertebrates, their clavicles are fused to form a furcula or wishbone, and they have a deeply keeled sternum for the attachment of large flight muscles. These and many other features particular to birds are related to flight and will be discussed in more detail in Chapter 2. Here in Chapter 1, I want not so much to consider what birds are, but rather to explore their evolutionary history.

And the answer to my opening question? Well I vividly recall my own incredulity as a student when my smiling tutor declared 'Birds are the last of the dinosaurs!'

Chapter overview

Concept
Synapsids vs diapsids

Reptiles are classified into one of two types depending upon the morphology of the skull, or more precisely upon the arrangement of fenestrae (windows) in the skull. Synapsid reptiles have a single fenestration behind the eye socket, whereas the skulls of diapsids have two, one above the other. As a result, diapsids have a lighter and generally more 'open' skull. They also tend to have a lighter skeleton and more slender body—they are more bird like.

1.1 Birds are dinosaurs

There are currently two main theories to explain the evolutionary origins of birds. Both agree that birds arose from reptiles, as of course did mammals. But whereas the mammals are thought to have their origins in the synapsid reptiles, the birds are thought to have evolved from the diapsid reptiles. The numerous species of lizard and snake living today are diapsids, but to uncover the evolutionary roots of birds we need to look back to the now extinct archosaurs, perhaps the most successful of the diapsid groups historically giving rise as it did to the thecodonts, the pterosaurs, and most notably the dinosaurs (Figure 1.1).

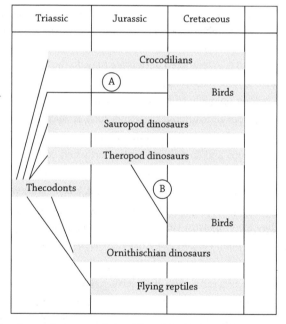

Figure 1.1 Putative evolutionary relationships of the birds and their closer relatives. One school of thought suggests that birds evolved from thecodont ancestors in the Triassic period (A). It seems to be far more likely, however, that they had a much later origin, diverging from the theropod dinosaurs in the mid Jurassic (B). Adapted from Gill, F.B. (2007) *Orinithology*, 3rd edn. Freeman, New York.

One school of thought suggests that the birds have their origins in the thecodonts and evolved some 230 million years ago (mya) at about the same time as the dinosaurs. This would mean that the birds were not so much dinosaurs, more their distant cousins. In contrast, the second school of thought argues that rather than evolving alongside the dinosaurs, birds evolved directly from one of the dinosaur sub-groups—the theropods—some 150 mya. The arguments for and against each scenario have been made forcefully, each school of thought putting forward sophisticated interpretations of the details of the fossil remains of various prehistoric animals that may be linked to the bird story. This is a complex and detailed specialist area and one that I do not wish to venture into in a book with a scope as wide as this one has. Instead I would recommend that the interested reader dip into one or other (or for balance preferably both) of the titles recommended as key texts. On balance, however, I feel that the weight of evidence seems currently to support the dinosaur (theropod) hypothesis.

Flight path: evolution of feathers. Chapter 2.

Key reference

Feduccia, A. (1996) *The Origin and Evolution of Birds.* Yale University Press, New Haven. Paul, G.S. (2002) *Dinosaurs of the Air: The Evolution and Loss of Flight in Dinosaurs and Birds.* Johns Hopkins University Press, Baltimore.

1.2 *Archaeopteryx*

In 1860, just 1 year after the publication of Charles Darwin's world-changing book *On the Origin of Species*, a fossil was discovered in the Solhnofen Lithographic Limestone of southern Germany that caused a sensation. The fossil was that of a single secondary flight feather and it was the first conclusive evidence of the existence of prehistoric birds. Within weeks, in 1861, a second fossil was announced from the same region—that of an almost complete feathered skeleton (this is now referred to as the London specimen because it resides in London at the Natural History Museum, Figure 1.2). A bird—but not just a bird, a bird with reptilian or dinosaur features—exactly the putative 'missing link' that Darwin had predicted in his book. This fossil was destined to become one of the most familiar and at the same time most fiercely debated in history.

From the handful of *Archaeopteryx* fossils that have been discovered we now know that it was feathered in a way consistent with flight and that it had a broadly bird-like skeleton (Figure 1.3). However, because it lacked a developed sternal keel we cannot be sure that it was capable of powered flight and so it may have been more of an accomplished glider. But then again it did have a reduced pelvis (like modern flying birds), an enlarged furcula and coracoid, and a broad sternum, all of which would have facilitated the attachment of larger breast muscles (essential for flight) and so perhaps it did fly in the true sense of the word. It does seem to have been at home on the ground and has relatively long/strong running legs—perhaps it was a wader, or darted through scrub like a modern Roadrunner *Geococcyx* sp. Significantly it had an arrangement of toes similar to that of modern perching birds, three of them pointing forwards and one pointing backwards. This backwards-pointing toe, which is properly termed the hallux, is not found in any

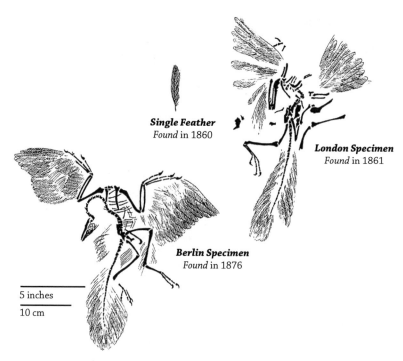

Single Feather
Found in 1860

London Specimen
Found in 1861

Berlin Specimen
Found in 1876

5 inches

10 cm

Figure 1.2 *Archaeopteryx* remains. The first feather, the London specimen, and the 1876 Berlin specimen. Adapted from Chatterjee, S. (1997) *The Rise of Birds: 225 Million Years of Evolution*. The Johns Hopkins University Press, Baltimore.

Key reference

Yalden, D.W. (1985) Forelimb function in *Archaeopteryx*. In: Hecht, M., Ostrom, J., Viohl, G., and Wellnhofer, P. (eds) *The Beginnings of Birds*. Freunde des Jura-museum, Eichstätt, pp. 91–97.

of the known non-avian dinosaurs. It had well-developed forelimbs and in fact these may have been more important in locomotion than the legs, being longer, stronger, and equipped with claws. Perhaps it swam like a young Hoatzin *Opistha-comus hoazin*. It had fingers suited to climbing and it had claws typical in curvature of modern perching birds (Figure 1.4). We know this because of the work of Derek Yalden and Alan Feduccia who have obtained curvature measurements of the claws of a large number of species of living bird and shown that they form three natural groupings: ground dwellers like the pheasants with relatively straight claws; tree climbers like the woodpeckers with strongly curved claws; and perching birds like the finches with intermediate claw curvature. The mean claw curvature value of the available *Archaeopteryx* specimens together with the presence of the backwards-pointing hallux makes it highly likely that this was a perching bird.

The teeth of *Archaeopteryx* suggest a carnivorous diet and it may be that the clawed hands were used to catch and grasp large invertebrates or small vertebrates—perhaps it ate fish. After all, paleoecological reconstructions of the conditions at the time that the fossil-bearing sediments were laid down suggest a habitat of scrubby islands in shallow lagoons. A predatory habit is further supported

Figure 1.3 The skeleton of *Archaeopteryx* (A) in comparison with that of a modern domestic pigeon (B). Note the expansion of the modern brain case (1), the fusion of the modern hand bones (the wing) (2), the fusion of the pelvic bones and reduction of the tail to form the modern pygostyle (3 and 4), the development of the sternum in the modern birds (to facilitate the attachment of large flight muscles (5), and the strength-ening of the modern rib cage (6). From Colbert, E.H. (1955) *Evolution of Vertebrates*. John Wiley and Sons Inc., New York.

by the suggestion that this was a large-eyed animal with a degree of binocular vision and a larger brain capacity than contemporary animals (although not yet on a par with modern birds).

Who knows, perhaps the next *Archaeopteryx* fossil to be described will help palaeontologists to firm up on some of this speculation.

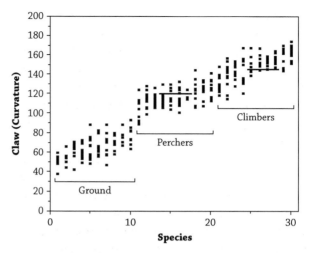

Figure 1.4 The variation in claw curvature in 30 species of modern bird, grouped according to lifestyle characteristics. The mean curvatures of the claws of fossilized *Archaeopteryx* foot claws are plotted as a solid line (amongst the perchers); those of the hand/wing are plotted as a solid line amongst the climbers. From Feduccia, A. (1993) Evidence from claw geometry indicating arboreal habits of Archaeopteryx. *Science* **259**, 790–793.

1.3 The evolution of modern birds

Although *Archaeopteryx* does have a significant role to play in the story of the evolution of birds, it is not in itself the ancestor of modern birds. It seems more likely that it represents a dead-end branch on the bird evolutionary tree; and one that died out before the end of the Jurassic period (145.5 mya). In fact, *Archaeopteryx* is just one of a growing number of prehistoric birds known only from fossil remains. Notable are the *Enantiornithes*, a diverse group of more than 25 known types of mostly flying birds dated to the early Cretaceous, the fish-eating and tooth-beaked *Ichthyornithiformes* (flying birds), and *Hesperornithiformes* (mostly swimming birds, although some of them did fly). We also have available a large number of modern birds known only from fossils and sub-fossils. The discovery and description of more and more fossils will undoubtedly help us to better understand this stage of the evolution of birds. But it would be wrong to assume that the existence of specimens of modern birds means that we fully understand their later evolution, and the relationships of many modern taxa are far from resolved.

When did the modern lineages arise?

A key question in the story of the evolution of birds relates to the extinction of the non-avian dinosaurs. Did modern birds coexist with non-avian dinosaurs?

Classification and the nomenclature of birds, an example The House Sparrow:	
Class	Aves
Sub-class	Neornithines
Superorder	Neognathae
Order	Passeriformes
Family	Passeridae
Genus	*Passer*
Species	*domesticus*

On the other hand, did the modern bird lineages evolve at some time later than the catastrophic extinction event that occurred at the Cretaceous/Tertiary (K/T) boundary approximately 65 mya? Two complementary lines of evidence are available to those who seek an answer to this question. Palaeontologists are able to date fossil remains through their association with sediment deposits of known age, i.e. they use a geological clock. Molecular biologists, on the other hand, are able to calibrate rates of gene mutation within organisms through evolutionary time, i.e. they use a molecular clock.

Palaeontologist Julia Clarke and colleagues have described a fossil bird collected from Vega Island in the western Antarctic; from their analysis they suggest that they have provided a definitive answer to this question. This important fossil is an incomplete specimen of a species that has been named *Vegavis iaai*. Although the examination of the detail of the bone structure of this bird suggests that it is sufficiently distinct to warrant the status of a new species, it has enough in common with other previously described fossils, and more crucially with extant forms, that it can be confidently placed within the modern Anseriformes. Specifically it is apparently closely related to the modern ducks and geese (the Anatoidea).

The fossil was found in sandstone deposits dated as coming from the middle/upper Maastrictian (the geological period defined as comprising those sediments laid down some 66–68 mya). This means that this bird was a contemporary of the last of the non-avian dinosaurs and strongly suggests that the evolution of the modern bird lineages occurred during the cretaceous period, i.e. modern birds really are dinosaurs that survived the K/T extinction event.

Similarly, molecular biologist Tara Paton and her colleagues have carried out an investigation the aim of which was to test a hypothesis proposed by Feduccia that all modern birds had evolved post-K/T extinction from a lineage of transitional shorebirds (one of few prehistoric bird lineages he proposed to have survived the event). By comparing the mitochondrial DNA (mtDNA) of various species of birds from a range of taxa they were able to demonstrate that many of the modern orders of bird were extant some time prior to the K/T boundary (Figure 1.5) and that modern birds were not therefore descended from a transitional shorebird. In fact their work suggests that the main divisions of modern birds, the paleognaths (ratites and tinamous) and the neognaths (all other modern birds), probably diverged from one another some 123 mya.

Key reference

Clarke, J.A., Tambussi, C.P., Noriega, J.I., Erikson, G.M., and Ketchan, R. (2005) Definitive fossil evidence for the extant avian radiation in the cretaceous. *Nature* **433**, 305–308

1.4 The phylogeny of birds

Morphological phylogeny

There is at present no single accepted phylogeny of the modern birds, that is to say we do not yet have a definitive statement of the evolutionary relationships of

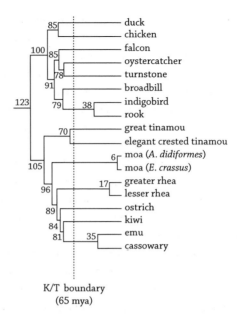

Figure 1.5 The estimated divergence times of the modern lineages of birds. Numbers printed at divergence points are millions of years before the present (mya). The K/T boundary (65 mya) denotes the extinction of the dinosaurs. Adapted from Paton, T., Haddrath, O., and Baker, A.J. (2002) Complete mitochondrial DNA genome sequences show that modern birds are not descended from transitional shorebirds. *Proceeding of the Royal Society B: Bological Sciences* **269**, 839–846.

different groupings of birds. Traditionally taxonomists have attempted to uncover evolutionary relationships through the comparison of morphological characters. Figure 1.3 provides an example of this kind of comparison; the bones highlighted are basic characters shared by both *Archaeopteryx* and the modern domestic pigeon. If we were to add to this comparison the skeleton of a Passenger Pigeon (*Ectopistes migratorius*), we would be able to use the closer similarities between the two pigeons than between either and *Archaeopteryx* to infer that the pigeons were the more closely related pair in an evolutionary sense.

Character conservation and convergence

However, it is not always the case that apparent similarity of shared characters equates to close evolutionary ties. The order Passeriformae comprises more than 5,000 species or more than half of all extant bird species. As such they can reasonably be described as being the dominant component of the world's avifauna. Passerines also occupy a bewildering array of ecological niches and exhibit an amazing range of morphologies. However, we can be confident that they form what is known as a monophyletic group, i.e. all of the current passerine species have evolved from a single common ancestor.

We know this because all passerine birds share features in common that are unique to the passerine order. For example, most birds have a preen gland (or uropygial gland) low on their back just above the base of the tail. This gland produces an oily secretion that is used by birds to maintain the physical quality of their feathers and to regulate bacterial and fungal communities that grow on them. In the case of aquatic birds the preen gland is particularly large and its secretions are important in feather water-proofing. Preen glands vary in their structure, but all passerines share a unique preen gland morphology. The passerines also all share a unique sperm morphology. The spermatozoa of the birds of most orders are straight whereas those of the passerines are helical in structure and move forwards via a spinning action. The passerine preen gland and sperm type are therefore conserved characters, i.e. they have evolved once in the early passerine ancestor and been retained throughout the passerine radiation.

Generally those characters that remain fixed in the face of ecological adaptation, so-called conservative characters, are the most useful in classification. Where characters do respond readily to ecological selection pressures convergence in evolution may lead to confusion in apparent relationships. For example, all passerine birds have a foot that is adapted for perching, having three forward-pointing toes and one backward-pointing toe. This toe arrangement is termed anisodactyly and is illustrated in Figure 1.6.

Anisodactyly is a conserved character that can be used to confirm membership of the passerine order (because it is only exhibited by birds that also exhibit all of the other uniquely passerine features). It has evolved once in the passerine ancestor and has been retained thereafter. Figure 1.6 also illustrates an alternative toe arrangement—zygodactyly. The zygodactyl foot does not indicate membership of a single order. In fact close examination of the precise arrangement of bones of the foot suggests that zygodactyly is an example of character convergence, i.e. it has evolved on a number of independent occasions in a range of unrelated species and orders including the osprey (*Pandion haliaetus*), the woodpeckers (Picidae), the owls (Strigidae), and some swifts (Apodidae).

Not all of the characters used in taxonomy are skeletal/anatomical however, and comparisons of plumage, behaviour, vocalizations, and even ectoparasites have all been used in attempts to elucidate avian evolutionary relationships. But perhaps the most exciting development in efforts to unravel the avian phylogenetic tree in recent times is the use of biomolecular information—information about genes and their chemical products.

Key reference

Livezey, B.C. and Zusi, R.L. (2007) Higher-order phylogeny of modern birds (Theropoda, Aves: Neornithes) based on comparative anatomy. II. Analysis and discussion. *Zoological Journal of the Linnean Society* **149**, 1–95

Biomolecular phylogeny

In 1990 Sibley and Ahlquist published a phylogeny of avian families based on the process of DNA–DNA hybridization. In simple terms they directly compared the chemical structure of the DNA of different bird species. Their work pioneered the use of biochemical characters in avian phylogeny, but their results did not produce

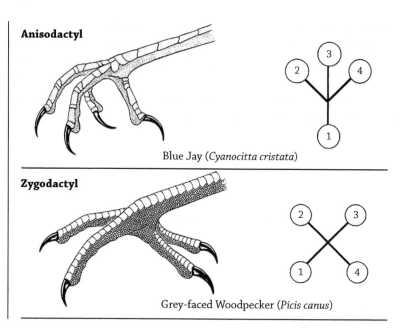

Anisodactyl

Blue Jay (*Cyanocitta cristata*)

Zygodactyl

Grey-faced Woodpecker (*Picis canus*)

Figure 1.6 Examples of the toe arrangements of two modern birds: the ansiodactyl Blue Jay *Cyanocitta cristata*, a perching bird (A); and the zygodactyl Grey-faced Woodpecker *Picis canus*, a climbing bird (B). These are just two of 15 different avian toe arrangements. Adapted from Proctor, N.S. and Lynch, P.J. (1993) *Manual of Ornithology: Avian Structure and Function*. Yale University Press, New Haven.

Concept
DNA deoxyribonucleic acid

The nuclei of our cells contain chromosomes; paired strands of DNA arranged in the iconic double helix with which we are all no doubt familiar.

Sequences of the basic units of these strands act as templates for the production of proteins— these templates are our **genes**. We inherit one strand of each of our chromosomes from our father and one from our mother.

Mitochondria, the power- houses of our cells, also each contain a small circle of DNA. This mtDNA we inherit only from our mothers.

When genes are inherited they may change, either as a result of the recombination of nuclear DNA from two parental sources, or as a result of **mutations**, acci- dental copying errors in either nuclear DNA or mtDNA.

Key reference
Biochemical phylogeny

Cracraft J. et al. (2004) Phylo- genetic relationships among modern birds (Neornithies). In: Cracraft J. and Donoghue M.J. (eds) *Assembling the Tree of Life*. Oxford University Press, New York.

a definitive avian tree and were limited because of the small units of DNA used and therefore the limited number of characters involved in tree construction. Subse- quent workers in the field have used improved molecular techniques to gain a better level of discrimination between taxa, particularly, but not exclusively, at the species level, through the comparison of mtDNA and nuclear gene sequences.

Putative phylogenies

I began this section by stating that there is no single accepted phylogeny of birds, and that is the case. However, I have decided to provide three examples of puta- tive phylogenies in Figures 1.7–1.9. Figure 1.7 illustrates the phylogeny of avian orders proposed by Livezey and Zusi. This is a morphological phylogeny based upon an analysis of a staggering 2954 characters. Figure 1.8 illustrates the phylo- genetic hypothesis of Cracraft and coworkers. This is a biochemical phylogeny of avian families arrived at as a summary of a number of available alternative molec- ular phylogenies. Figure 1.9 is a phylogeny of the families of the order Passeridae proposed by Barker and coworkers, who have compared the sequences of nucleic acids of two nuclear genes, *RAG-1* and *c-mos*, in a range of passerine families.

Their phylogenetic hypothesis is significant because it suggests an alternative to the previously held view that the passerines of the Australian region represent an endemic radiation from a few ancestral colonist species. The basal position of the New Zealand wrens (Acanthisittidae) and other passerine families found only in that region suggests that the passerines first evolved here and subsequently radiated outwards to colonize the rest of the world.

Key reference

Barker, F.K., Barrowclough, G.F., and Groth, J.G. (2002) A phylogenetic hypothesis for passerine birds: taxonomic and biogeographic implications for an analysis of nuclear DNA sequence data. *Proceedings of the Royal Society B: Biological Sciences* **269**, 295–308.

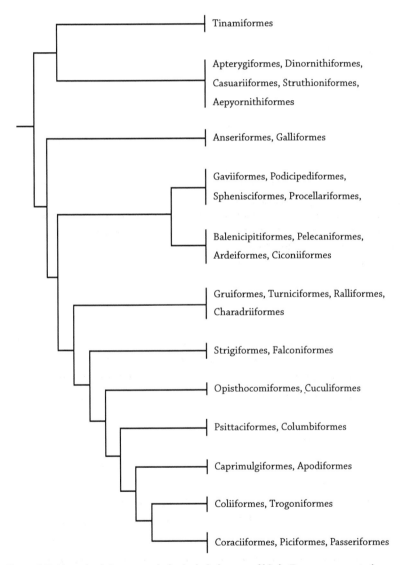

Figure 1.7 Hypothesis for a morphological phylogeny of birds. Taxa are presented as orders. For more familiar names refer to the appendix at the end of this chapter. Adapted from Livezey, B.C. and Zusi, R.L. (2007) Higher order phylogeny of modern birds (Theropoda, Aves: Neornithes) based on comparative anatomy. II. Analysis and discussion. *Zoological Journal of the Linnean Society* **149**, 1–95.

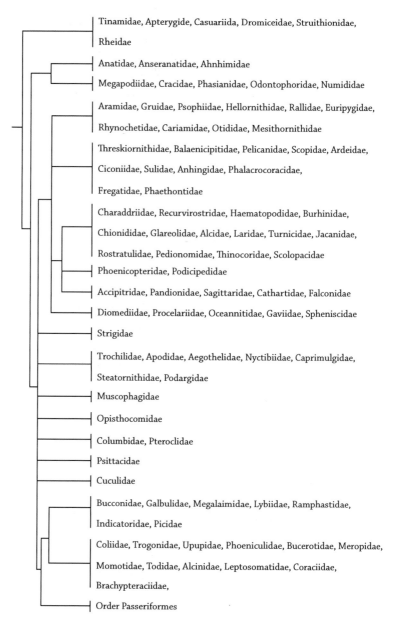

Figure 1.8 Hypothesis for a molecular phylogeny of birds. Note that whilst the majority of taxa are given at family level, the many families of the order Passeriformes are grouped for clarity. For more familiar names refer to the appendix at the end of this chapter. Adapted from Cracraft, J. *et al.* (2004) Phylogenetic relationships among modern birds (Neornithies). In: Cracraft J. and Donoghue M.J. (eds) *Assembling the Tree of Life.* Oxford University Press, New York, pp. 468–489.

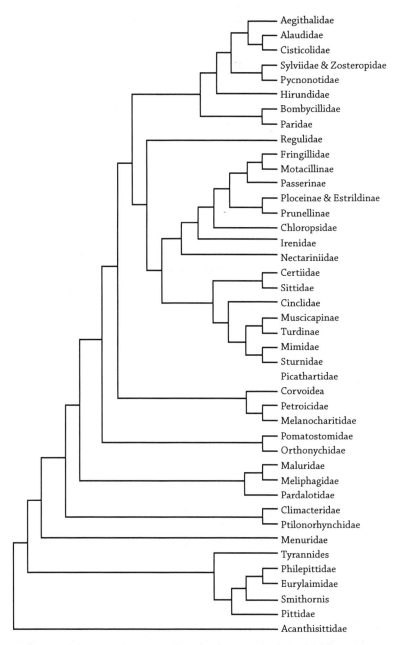

Figure 1.9 Hypothesis for a molecular phylogeny of passerine bird families. For more familiar names refer to the appendix at the end of this chapter. Adapted from Barker, F.K., Barrowclough, G.F., and Groth, J.G. (2002) A phylogenetic hypothesis for passerine birds: taxonomic and biogeographic implications for an analysis of nuclear DNA sequence data. *Proceedings of the Royal Society B: Biologicval Sciences* **269**, 295–308.

In their excellent essay 'Phylogenetic relationships among modern birds (Neornithies)' (see Figure 1.8), Cracraft and colleagues state that *'These are exciting and productive times for avian sytematists.'* They go on to discuss the potential that advances in molecular biology provide for the elucidation of the avian phylogeny, particularly when integrated analyses combining biochemical and morphological characters are employed. Perhaps the avian tree of life is within our grasp.

1.5 Adaptive radiation and speciation

One of the first things that a bird watcher notices is that there is an amazing diversity of form even within groups of closely related bird species. Living, as I do, on the coast, the bills of the wading birds are my favourite example of this phenomenon. I regularly see as many as a dozen species feeding along the same area of shore, all of them members of the same order, the Charidriiformes, but ranging in bill morphology from the tiny straight bill of the Little Ringed Plover *Charidrius dubius* used delicately to pick isopods from the strand line, to the long de-curved bill of the Eurasian Curlew *Numenius arquata* used to probe deep into wet sand to drag out large polychaetes, to the heavy hammering bill of the Eurasian Oystercatcher *Haematopus ostralegus* ideal for smashing open the shells of bivalve molluscs (Figure 1.10).

This diversity of bill shape allows the birds to coexist without competing too closely for food—they each occupy their own feeding niche. As such their bills and

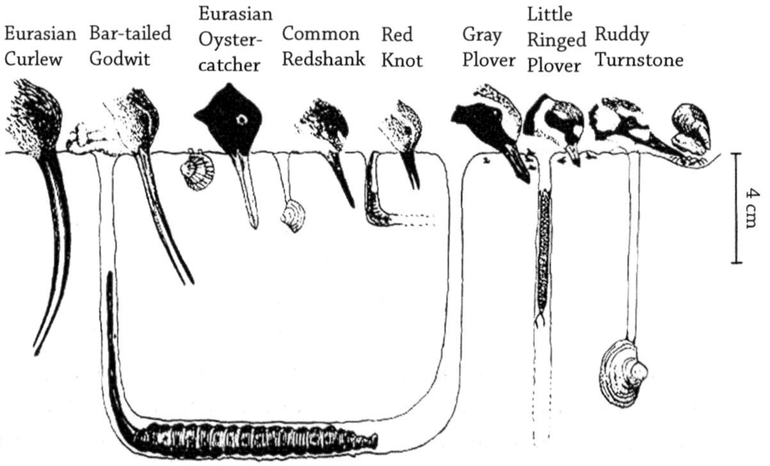

> **Flight path:** foraging behaviour and the concept of niche. Chapter 6.

Figure 1.10 Because their bill morphology varies, species of wading shore birds are able to forage alongside one another with minimal competition. From Gill, F.B. (2007) *Orinithology*, 3rd edn. Freeman, New York (after Goss-Custard, J.D. (1975) Beach feast. *Birds* September/October, 23–26).

associated mode of feeding can be described as evolutionary adaptations which enhance their individual survival by reducing the interspecific competition that they experience.

The observed diversity in bill morphology is one visible outcome of a process of adaptive radiation—we can presume that at one time the ancestral Charidriiform bill was pretty uniform, but that through evolutionary time and in parallel with the evolution of all of the various wader species it has evolved as a result of natural selection.

Darwin's finches

This is the same phenomenon that has given rise to perhaps the most iconic of examples of adaptive radiation, that of the finches of the Galapagos and Cocos Islands (Figure 1.11). There are 15 species of so-called Galapagos finch; 14 of them are endemic to the Galapaogs islands and one is only found on the Cocos islands. Galapagos, or Darwin's, finches, as they are also often called (Charles Darwin did collect the first specimens of these species, but they were not quite as pivotal in the development of his ideas as is often assumed to be the case), have been presumed to have evolved from an ancestral species that accidentally colonized these isolated islands from the South American mainland close to 1,000 km away. Through the comparison of nuclear DNA and mtDNA sequences of the Galapagos finches and a range of putative mainland relatives, Sato and colleagues have recently deduced that the closest living mainland relative of the Galapagos finches is the Dull-coloured Grassquit *Tiaris obscura*, a species which inhabits humid forest edges, farmland, and scrub throughout Venezuela, Colombia, western Ecuador, and western and southern Peru, habitats not dissimilar to those encountered in the islands by the colonists. *T. obscura* is not the direct ancestor of the Galapagos finches, but it is likely that they shared a common ancestor.

The early colonists probably arrived on just one of the islands and found an environment devoid of competition but rich in potential resources. They are presumed to have formed a self-sustaining population from which groups of individuals went on eventually to colonize the other islands in the group. Living on these isolated islands, each population of birds may have become a specialist in the exploitation of a particular resource and through the process of natural selection (see Box 1.1) their populations will have diverged from one another in terms of ecology, behaviour, and, crucially, genetics. Through time these isolated populations of birds may have diverged from one another to such a level that were two populations to re-colonize the same island and coexist they would remain isolated from one another in the sense that they would have evolved to become separate species.

The coexistence of two or more finch species on a single island would probably have increased the rate of divergence as birds increased their levels of behavioural/

Key reference

Sato, A., Tichy, H., O'hUigin, C., Grant, P.R., Grant, B.R., and Klein, J. (2001) On the origin of Darwin's finches. *Molecular Biology and Evolution* **18**, 299–311.

Concept
Species

The species is the unit we use to measure biological diversity. Typically species are defined as populations of potentially interbreeding individuals that are reproductively isolated from one another (Biological Species Concept).

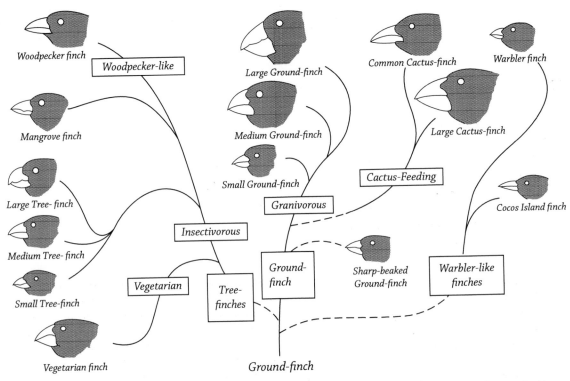

Figure 1.11 The adaptive radiation of Darwin's finches. From a single ancestor the process of speciation has resulted in the evolution of 15 species of finch forming three distinct ecological groups: Ground-finches specializing in seeds of various sizes; Warbler-like finches which use a wide range of feeding techniques; and Tree-finches, some species of which have evolved tool use—using cactus spines to pry prey from crevices. From Lack, D. (1947) Darwin's Finches. Cambridge University Press.

ecological isolation from one another as they became increasingly specialized to avoid interspecific competition.

Hybrids

When the behavioural and genetic isolation between two species is incomplete it may be possible for hybridization to occur. In some cases these hybrids might be sterile and so make no genetic contribution to the next generation themselves; as such they are an evolutionary dead-end and are probably regarded as no more than a curiosity. But occasionally these hybrids are fertile and may reproduce either amongst themselves or with one or other of their parent species. There are two main outcomes possible as a result of this phenomenon. In some cases hybrid zones are formed when the stable ranges of two geographically adjacent species

Box 1.1   Evolution in action: natural selection and the morphology of finch bills

I am sure that if you take the time to think about the characteristics of populations you will note the following:

- Not all individuals are the same. That is to say there is considerable variation within a species.
- Offspring resemble their parents. This is because heritable variations are passed from parents to their offspring through their genes.
- Reproductive overproduction is common. Many more individuals are produced than can ever survive to mature and reproduce.

These basic observations, a prodigious amount of patient work, and probably a touch of genius, allowed Charles Darwin to construct his theory of evolution by natural selection. Long before the importance of genes was appreciated he realized that some heritable property of certain individuals placed them at an advantage relative to others of their kind. Through time, the differential survival of these individuals should result in a shift in the make-up of the population such that animals without the advantage become increasingly rare (or perhaps disappear) and those possessing it become more common. He proposed a scenario under which beneficial traits were selected for and detrimental traits were selected against.

We now understand that selection can be strongly directional in the way that I have just described, leading to the increasing dominance of a trait or to its eradication, or it can act in a disruptive way resulting in the evolution of two populations that have very different but successful phenotypes. Selection can also act in a stabilizing way to maintain the status quo (Figure 1.12).

So can we see evolution by natural selection actually happening? Well thanks to the work of Peter Grant, Peter Boag, and their colleagues we can. They have carried out a very detailed study of the population of the Medium Ground Finch *Geospiza fortis* on the tiny island of Daphne Major, one of the Galapagos Islands. Since the mid 1970s these researchers have

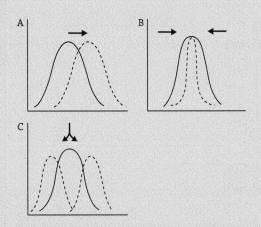

Figure 1.12 The effect of directional (A), stabilizing (B), and disruptive (C) selection upon a hypothetical population. In each case an arrow indicates the direction of the selection taking place, the solid line describes the preselection population and the broken line the post selection population(s). From Scott, G.W. (2005) *Essential Animal Behavior*. Blackwell Science, Oxford.

ringed and measured almost all of the *G. fortis* on the island, they have recorded year on year survivorship, and have regularly trapped birds to take measurements from them. These measurements include mass, wing length, tarsus length, and bill length and depth. During the early years of their study, Daphne Major received regular seasonal rainfall (around 130 mm per year) and the finch populations did well. But in 1977 the island suffered an intense drought with just 24 mm of rain falling during the wet season. The impact upon the birds was dramatic. They made no breeding attempts, many birds delayed their moult, or failed to moult at all, and significant levels of mortality were recorded (an 85% decline in the population in just one season).

The drought did not just affect the birds on the island. Many plants failed to set seed, leading to a food shortage for the finches, and most of the

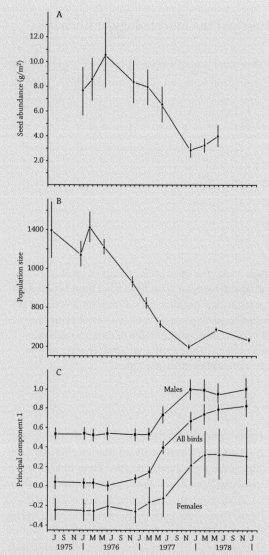

finch mortality could be attributed either directly or indirectly to starvation. However, some birds did survive. They tended to be the bigger birds in the population with bigger beaks (Figure 1.13).

In effect what has happened here is a burst of intense directional selection. Initially the population exhibited size variation—smaller birds with smaller beaks and bigger birds with bigger beaks. When the seed crop failed the available seeds were used up quickly, and initially it seems that smaller seeds were consumed preferentially. When seeds became scarce the bigger birds changed their feeding behaviour to exploit the bigger seeds, but the smaller birds did not have the beak morphology needed to crack these seeds, and starved. Thus some of the observed variation in beak size was lost from the population, which became dominated by bigger birds. But surely that wasn't the first drought to have occurred so why were there small birds in the population? By continuing to study the finch population of Daphne

Figure 1.13 Under drought conditions in 1977/78 seed production on the Galapagos Islands fell (A). Without food, the finch population crashed (B) but some birds did survive. The principle component plotted in (C) is a mathematical measure of the morphology of the finch population. As the drought progresses, morphology clearly changes. In fact, by 1978 only larger birds with larger beaks remain in the population. Adapted from Boag, P. and Grant, P. (1981) Intense natural selection in a population of Darwin's finches (Geospizinae) in the Galapagos. *Science* **214**, 82–85.

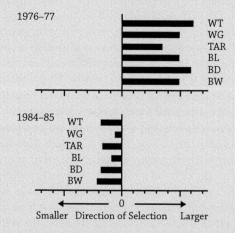

Figure 1.14 Oscillating selection for finch size in response to climate. During dryer periods (such as the mid 1970s), selection favours bigger birds. However, following wet periods (such as the El-Niño of 1982/83), smaller size is favoured. Biometrics listed include: WT, weight: WG, wing length; TAR, tarsus length; BL, bill length; BD, bill depth; BW, bill width. Adapted from Gibbs, L.H. and Grant, P. (1987) Oscillating selection on Darwin's finches. *Nature* **327**, 511–13.

Major these dedicated researchers seem to have the answer. Although natural selection favours large size in response to drought, small size is favoured in response to particularly wet years (when there are many more smaller seeds than larger seeds) and during the first year of a bird's life (possibly for metabolic reasons). It would seem therefore that, in effect, size in *G. fortis* is an example of oscillating selection, i.e. repeated bursts of opposing directional selection in response to extremes of climate and associated fluctuations in food supply (Figure 1.14). This work demonstrates two important things: it shows us that we can see evolution in action and it shows us that very long-term studies are important.

overlap. In other cases transient hybridization, or genetic swamping, occurs as the range of one species expands and previously isolated species are temporarily brought into contact with one another, the usual outcome being the disappearance of one of the species.

The classic example of the stable hybrid zone is that of the black and grey Hooded Crow (*Corvus corone cornix*) of northern and eastern Europe and the all black Carrion Crow (*Corvus corone corone*) of southern Europe. (Note that each of these birds has a latin name in three rather than the usual two parts. This is because they are not considered by taxonomists to be sufficiently distinct to be given full species status; the third name denotes that they are different as sub-species.) Along the narrow (2,100 km long but only 50–120 km wide) contact zone between the populations of these two crows they freely interbreed and their hybrids exhibit a plumage that is intermediate between the two types (black with varying amounts of grey).

One fascinating example of transient hybridization is that of the Blue-winged and Golden-winged Warblers (*Vermivora pinus* and *V. chrysoptera*) of North America. About 200 years ago ornithologists would have known the Blue-winged Warbler as a southern species and the Golden-winged Warbler as a species found in the north, but today the ranges of the two species overlap and that of the Blue-winged Warbler is increasing whilst the Golden-winged Warbler is becoming increasingly scarce. This shift in the range of the Blue-winged Warbler is a response to an increase in their preferred secondary scrub habitat and is largely a result of human clearances of forested areas. Increased scrub cover also benefited Golden-winged Warblers initially but where the two species do overlap Golden-wings eventually disappear. This disappearance takes about 50 years so it isn't simply the case that the Blue-wings see off the competition immediately. In fact the two species do coexist initially but where they do two new 'species' also appear. These are hybrids. One, Brewsters Warbler (*Vermivora chrysoptera* × *pinus*, or *V.* 'leucobranchialis'), is a bird with the body plumage of a Golden-winged Warbler and the face pattern of a Blue-winged. The other is Lawrence's Warbler (*Vermivora pinus* × *chrysoptera* or *V. lawrencei*') which has a Golden-wing face pattern and Blue-wing body (Figure 1.15).

The pattern of the hybrids' appearance is predictable whenever the two parent species come into contact. Initially there are good numbers of Golden-wings in a population, but as numbers of immigrant Blue-wings increase the numbers of

Lawrence's Warbler
(*Vermivora pinus x chrysoptera,*
or *V. 'lawrencei'*)

Brewster's Warbler
(*Vermivora chrysoptera x pinus,*
or *V. 'leucobronchialis'*)

Blue-winged Warbler
(*Vermivora pinus*)

Golden-winged Warbler
(*Vermivora chrysoptera*)

Figure 1.15 The Blue-winged Warbler/Golden-winged Warbler hybrid types. From Proctor, N.S. and Lynch, P.J. (1993) *Manual of Ornithology: Avian Structure and Function.* Yale University Press, New Haven.

Box 1.2   Hybridization and duck conservation

The White-headed Duck (*Oxyura leucocephala*) is an eastern European and central Asian species with a small isolated population in Spain. Populations are fragmented and under increasing pressure due to loss of habitat. This, coupled with the fact that populations are small and declining, makes this species globally endangered (the Spanish population fell to just 22 birds in 1977, and the main central Asian population fell from 100,000 in the 1930s to 10,000 in 2000). In Europe the species faces another

threat—genetic swamping through hybridization with an alien invader!

The alien in question is the Ruddy Duck (*Oxyuria jamicensis*), a close relative of the White-headed Duck but one that would naturally be isolated from it by the Atlantic Ocean. In their native range (North and South America) Ruddy Duck are a successful species with an increasing population (of more than half a million birds). They are an attractive duck and have been popular additions to wildfowl collections for many years. In Britain in the 1950s or early 1960s Ruddy Duck from such collections were accidentally released and a feral population quickly established itself such that the population grew from a handful of original birds to some 6,000 individuals in 50 years.

If the British Ruddy Duck population had stayed in Britain it may not have posed much of a problem but, as it grew, increasing numbers of Ruddy Duck presumed to originate from Britain were reported from a number of European countries. In Spain where invading Ruddy Duck and White-headed Duck came into contact, hybrids were quickly recorded. The hybrids were fertile and, second-generation hybridization was observed. In this situation there is obviously the possibility that the genome of the White-headed Duck in Spain would be swamped by that of the Ruddy Duck and another valuable population would disappear. And if the Ruddy Duck were to spread across Europe and invade Central Asia the White-headed Duck might disappear as a distinct species.

Extinction by genetic swamping might seem far fetched, but Judith Mank and others have demonstrated that in the case of at least one other duck species this might be just what is happening. Mank and colleagues have compared the genomes of American Black Duck (*Anas rubripes*) and European Mallard (*Anas platyrhynchos*). Mallard were introduced to the Americas probably by early settlers and they did very well there. They are closely related to Black Duck and so it is not perhaps very surprising that Mank and her colleagues found genetic similarities between them. But as well as making a comparison of today's populations she also took material from museum specimens collected between 1900 and 1935, and was therefore able to establish their contemporary and historical relatedness. Her results showed that the genetic distinction between the two species is shrinking, presumably as a result of continuing hybridization and genetic swamping. In the conclusion to her paper, Mank makes the somewhat sombre statement that '*The implications of our findings for the conservation of the black duck are grim. Without preventing hybridization, conservation of pristine black duck habitat will be ineffective in preserving the species*'. So can hybridization be prevented in such a situation?

In the case of the White-headed Duck situation it may not be too late. There are various articles of legislation that are used by the international community to bring pressure upon nation states to take action for conservation. In this case the relevant agreement is the Bonn Convention: the Convention on the Conservation of Migratory Species. Within the convention there is an agreement, The African–Eurasian Migratory Water bird Agreement (AEWA) designed to provide a legal framework for the conservation and management of 172 species of bird that are ecologically dependent upon wetland habitats. There are 116 signatory nations to the AEWA in Europe, Africa, north-east Arctic Canada, Greenland, Asia Minor, the Middle East, Kazakhstan, Turkmenistan, and Uzbekistan. The list includes a number of states which have White-headed Duck populations. Amongst other conservation measures the AEWA encourages states to assess the impact that alien species might have in the context of wetlands, and Article III of the agreement requires signatories to prohibit the deliberate introduction of non-native water birds into the wild and to take all reasonable measures to prevent their accidental release. This is very positive, but these steps will not solve the Ruddy Duck problem—the birds are already out there. However, the AEWA also requires that signatories '*ensure that when non-native species or hybrids thereof have already been introduced into their territory, those species or their hybrids do not pose a potential hazard to the population listed*'. So here is the legal requirement for the control of the Ruddy Duck in Europe. If that was insufficient in itself, the Council of Europe have published a specific action plan for the conservation of White-headed Duck, and the Bern Convention has put forward a strategy for the eradication of the Ruddy Duck throughout the Western-Palaearctic region. Ruddy Duck control measures are now in place, birds are culled throughout Europe, and early results are positive. There is some optimism that this particular threat to the White-headed Duck can be overcome. But of course the future security of the species will depend upon other conservation actions to protect specific habitats and populations.

References and further reading:
Hughes, B., Henderson, I., and Robertson, P. (2006) Conservation of the globally threatened white-headed duck, *Oxyura leucocephala* in the face of hybridization with the North American ruddy duck, *Oxyura jamaicensis*: results of a control trial. *Acta Zoologica Sinica* **52**, 576–578.

Mank, J.E., Carlson J.E., and Brittingham, M.C. (2004) A century of hybridization: decreasing genetic distance between American black ducks and mallards. *Conservation Genetics* **5**, 395–403.

Rehfisch, M.M., Blair, M.J., McKay, H., and Musgrove, A.J. (2004) The impact and status of introduced waterbirds in Africa, Asia Minor, Europe and the Middle East. *Acta Zoologica Sinica* **52**, 572–575.

Brewster's type hybrids also increase. These hybrids are fertile and backcross with both of the parent species, resulting in intermediate types and in the rarer Lawrence's type hybrid. As this process of hybridization and backcrossing continues Golden-winged Warblers become increasingly rare. Eventually Blue-winged Warblers dominate the community, Golden-winged Warblers have all but gone and hybrid types are rarely seen. However, as a result of the backcrossing of hybrids the genes of the Golden-winged Warblers may persist in the Blue-winged Warbler population, with the result that an occasional aberrant form may appear. This process of genetic takeover is referred to as 'swamping'.

Summary

Birds are specialist vertebrates thought to have evolved from theropod dinosaurs. The evolutionary relationships of modern birds species are not yet fully understood, but advances in phylogeny and new molecular techniques are bringing them within our grasp. Like all organisms birds continue to adapt and evolve in the face of environmental pressure.

Questions for discussion

1. When did birds evolve and what are their origins?
2. Is it ethically right to cull one bird species in order to conserve another?

Appendix 1

Familiar names of the members of the Orders and Families of modern birds listed in Figures 1.7, 1.8 and 1.9. This is not, however, an exhaustive list of bird orders/

families, and in this fast-moving field it should not be considered to be a definitive statement.

Tinamiformes	Tinamidae	Tinamous
Apterygiformes	Apterygidae	Kiwis
Dinornithiformes		Moas (extinct)
Casuariiformes	Casuariidae	Cassowaries
	Dromiceidae	Emu
Struthioniformes	Struthionidae	Ostrich
	Rheidae	Rheas
Aepyornithiformes		Elephantbirds (extinct)
Anseriformes	Anatidae	Ducks, geese, and swans
	Anseranatidae	Magpie goose
	Ahnhimidae	Screamers
Galliformes	Megapodiidae	Megapodes
	Cracidae	Curassows, guans, and chachalas
	Phasianidae	Pheasants, partridges, grouse, Turkeys, old world quail
	Odontophoridae	New world quail
	Numididae	Guineafowl
Gaviiformes	Gaviidae	Divers or loons
Podicipediformes	Podicipedidae	Grebes
Phoenicopteriformes	Phoenicopteridae	Flamingoes
Sphenisciformes	Spheniscidae	Penguins
Procellariformes	Diomediidae	Albatrosses
	Procelariidae	Shearwaters and petrels
	Oceannitidae	Storm-petrels
Balenicipitiformes	Balaenicipitidae	Shoebill
Pelecaniformes	Pelicanidae	Pelicans
	Sulidae	Boobies and gannets
	Anhingidae	Anhingas
	Phalacrocoracidae	Cormorants
	Fregatidae	Frigatebirds
	Phaethontidae	Tropicbirds
Ardeiformes	Ardeidae	Herons, egrets, and bitterns
Ciconiiformes	Threskiornithidae	Ibises and spoonbills
	Scopidae	Hamerkop
	Ciconiidae	Storks
Gruiformes	Aramidae	Limpkin
	Gruidae	Cranes
	Psophiidae	Trumpeters

	Hellornithidae	Finfoots
	Rallidae	Rails, gallinules, and coots
	Euripygidae	Sunbittern
	Rhynochetidae	Kagu
	Cariamidae	Seriemas
	Otididae	Bustards
	Mesithornithidae	Mesites
Turniciformes	Turnicidae	Button quails
Ralliformes	Rallidae	Rails
Charadriiformes	Charaddriidae	Plovers
	Recurvirostridae	Avocets and stilts
	Haematopodidae	Oystercatchers
	Burhinidae	Thick-knees
	Chionididae	Sheathbill
	Glareolidae	Pratincoles
	Alcidae	Auks
	Laridae	Gulls and terns
	Turnicidae	Button-quails
	Jacanidae	Jacanas
	Rostratulidae	Painted-snipe
	Pedionomidae	Plains-wanderer
	Thinocoridae	Seedsnipe
	Scolopacidae	Sandpipers
Strigiformes	Strigidae	Owls
Falconiformes	Accipitridae	Hawks, eagles, kites, and old world vultures
	Pandionidae	Osprey
	Sagittaridae	Secretary-bird
	Cathartidae	New world vultures
	Falconidae	Falcons
Opisthocomiformes	Opisthocomidae	Hoatzin
Cuculiformes	Cuculidae	Cuckoos, roadrunners, and anis
Psittaciformes	Psittacidae	Parrots, parakeets, macaws, and lories
Columbiformes	Columbidae	Pigeons and doves
Pterocidiformes	Pteroclididae	Sandgrouse
Caprimulgiformes	Nyctibiidae	Potoos
	Caprimulgidae	Nightjars or goatsuckers
	Aegothelidae	Owlet-nightjars
	Steatornithidae	Oilbirds
	Podargidae	Frogmouths
Apodiformes	Trochilidae	Hummingbirds
	Apodidae	Swifts

Coliiformes	Coliidae	Mousebirds
	Muscophagidae	Touracos
Trogoniformes	Trogonidae	Trogons
Coraciiformes	Bucerotidae	Hornbills
	Meropidae	Bee-eaters
	Momotidae	Motmots
	Todidae	Todies
	Leptosomatidae	Cuckoo-roller
	Alcinidae	Kingfishers
	Coraciidae	Rollers
	Brachypteraciidae	Ground-rollers
	Upupidae	Hoopoes
	Phoeniculidae	Woodhoopoes
Piciformes	Bucconidae	Puffbirds
	Galbulidae	Jacamars
	Megalaimidae	Asian barbets
	Lybiidae	African barbets
	Ramphastidae	Toucans and new world barbets
	Indicatoridae	Honeyguides
	Picidae	Woodpeckers and allies
Passeriformes	Acanthisittidae	New Zealand wrens
	Pittidae	Pittas
	Eurylamidae	Broadbills
	Philepittidae	Asites and false sunbirds
	Tyrannides	New world flycatchers
	Menuridae	Lyrebirds
	Ptilonorhynchidae	Bowerbirds
	Climacteridae	Australasian treecreepers
	Pardalotidae	Australo-Papuan warblers
	Meliphagidae	Honeyeaters
	Maluridae	Fairywrens
	Orthonychidae	Logrunners
	Pomatostomidae	Pseudo-babblers
	Melanocharitidae	Berrypeckers and longbills
	Petroicidae	Australasian robins
	Corvoidea	A paraphyletic group including crows, shirikes, drongos,jays, and others
	Picathartidae	Rock-fowl
	Sturnidae	Starlings
	Mimidae	Mockingbirds and thrashers
	Turdinae	Thrushes

Muscicapinae	Old world flycatchers
Cinclidae	Dippers
Sittidae	Nuthatches
Certiidae	Creepers
Nectariniidae	Sunbirds
Irenidae	Fairy-bluebirds
Chloropsidae	Leafbirds
Prunellinae	Accentors
Estrildinae	Waxbills and whydahs
Ploceinae	Weavers
Passerinae	Old world sparrows
Motacillinae	Wagtails and pipits
Fringillidae	Finches and Hawaiian honeycreepers
Regulidae	Kinglets
Paridae	Titmice and chickadees
Bombycillidae	Waxwings
Hirundidae	Swallows
Pycnonotidae	Bulbuls
Zosteropidae	White-eyes
Sylviidae	Old world warblers and gnatcatchers
Cisticolidae	Cisticolas
Alaudidae	Larks
Aegithalidae	Bushtit and long-tailed tits

CHAPTER 2

Feathers and flight

Birds are pilot and aircraft in one.

John Videler, 2006

No bird soars too high if he soars on his own wings.

William Blake, 1793

Although not unique to birds, the power of flight and feathers are probably their distinguishing feature in the eyes of most people. In this chapter I want to take some time to consider the feathers that make flight possible, their growth, their mainte-nance, and their replacement through moult. I will also describe the anatomical adaptations of birds to flight, and the process of flight itself.

Chapter overview

2.1 **Feathers**
2.2 **Feather tracts**
2.3 **Feather colour**
2.4 **Feather damage**
2.5 **Feather maintenance**
2.6 **Moult**
2.7 **Flight**
2.8 **The evolution of flight and flightlessness**

2.1   Feathers

Feathers were once thought to be one of, if not the, defining character of the birds, but recent discoveries of fossil dinosaur feathers indistinguishable from those of modern birds prove that this is not the case. It has also been a commonly held belief that feathers evolved in association with the evolution of flight, but this too can be discounted. Feathers of the modern type have been found on fossils of non-flying dinosaurs, including the ancestors of *Tyrannosaurs rex*. There is no doubt that feather types have evolved to make flying more efficient, but their original function must have been very different. Among modern birds, feathers are essential in flight and for water-proofing and insulation. They often have a role in communication, for courtship or competition, and in predator avoidance through camouflage. In some cases they even have a tactile function, an example being the feather-derived bristles around the mouths of some insectivores and some nocturnal birds.

> **Flight path:** evolution of birds from dinosaurs. Chapter 1.

Feather types

I am sure that you have in your mind a picture of the typical feather (perhaps something like the quill of the medieval scribe?). But if you took some time to think about all of the feather types that you have ever experienced their variety might start to bewilder you; ranging as they do from the simple bristles around the beak of some flycatchers to the magnificently ornate tail plumes of male Indian Peafowl *Pavo cristatus* (Plate 1). Thankfully for the purpose of this book we can reduce this bewildering array to essentially two basic types: contour feathers and downs.

Contour feathers

The contour feathers are all of those feathers that form the outline of the bird. They therefore include the tail feathers (reticies), the feathers of the wing (remiges), those covering the body of the bird, and the highly modified bristles that are often found around the head of the bird and may bear a superficial resemblance to mammalian hairs. The arrangement and extent of these various feather types are illustrated in Figure 2.1, and a range of typical feather types and structures are illustrated in Figure 2.2.

The typical contour feather has at its base a bare quill, or calamus. This is the part of the feather that sits inside the feather follicle and is attached to the body of the bird. The calamus extends into the often long and tapering central shaft of the feather which is more properly termed the rachis. In the case of most contour feathers (but not the bristles) the rachis supports two vanes, the blades of the feather. The vanes are actually composed of two opposite rows of barbs, projections of the rachis which in their turn support two parallel rows of smaller projections or barbules.

The barbules themselves are sculptured to allow them to lock together like Velcro, giving the vane its sheet-like quality. Specifically, those barbules which project from the barb and point forwards towards the tip of the feather (distal barbules) have a comb-like arrangement of hooked barbicels which lock into ridges on the barbicels on the adjacent backwards-pointing barbules (proximal barbules). Most feathers have a basal area of the vane (i.e. the area closest to the calamus) the barbs of which lack barbicels, resulting in a more open structure that can only really be

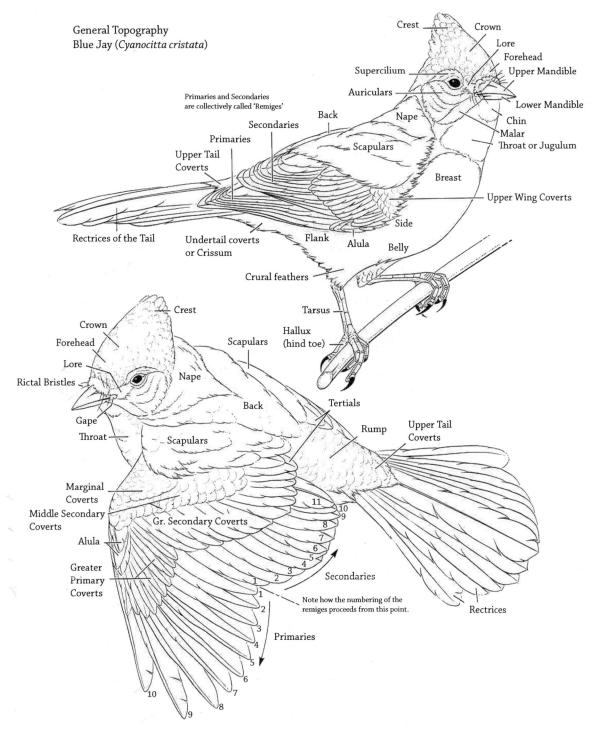

General Topography
Blue Jay (*Cyanocitta cristata*)

Crest

Crown

Lore

Forehead

Upper Mandible

Supercilium

Auriculars

Lower Mandible

Chin

Malar

Throat or Jugulum

Primaries and Secondaries
are collectively called 'Remiges'

Back

Nape

Secondaries

Primaries

Scapulars

Upper Tail
Coverts

Breast

Upper Wing Coverts

Side

Rectrices of the Tail

Undertail coverts
or Crissum

Flank

Alula

Belly

Crural feathers

Tarsus

Crest

Hallux
(hind toe)

Crown

Scapulars

Forehead

Lore

Nape

Rictal Bristles

Tertials

Back

Rump

Upper Tail
Coverts

Gape

Throat

Scapulars

Scapulars

Marginal
Coverts

Middle Secondary
Coverts

11

10
9

Gr. Secondary Coverts

8

Alula

7

6

5

4

Greater
Primary
Coverts

3

1

2

1

Secondaries

Note how the numbering of the
remiges proceeds from this point.

Rectrices

2

3

4

Primaries

5

6

10

7

8

9

Figure 2.1 The general topography of a bird. The species illustrated is the Blue Jay *Cyanocitta cristata*. From Proctor, N.S. and
Lynch, P.J. (1993) *Manual of Ornithology: Avian Structure and Function.* Yale University Press, New Haven.

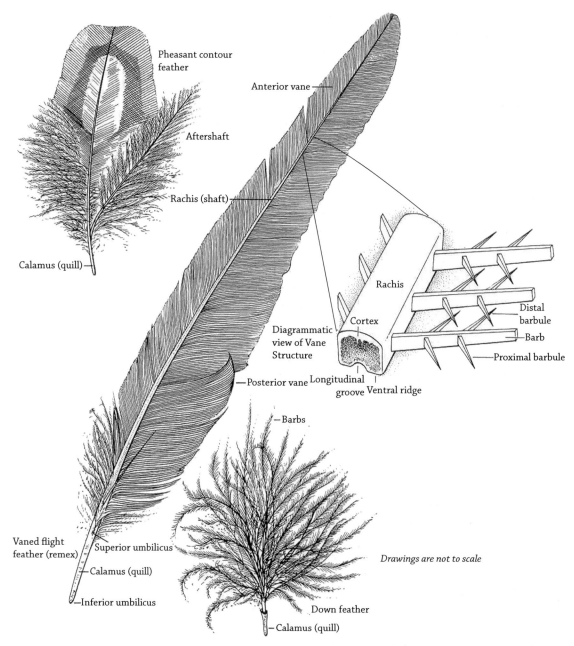

Figure 2.2 Feather structure. Adapted from Proctor, N.S. and Lynch, P.J. (1993) *Manual of Ornithology: Avian Structure and Function*. Yale University Press, New Haven.

described as 'fluffy'. A vane or vane area with this open structure is properly referred to as being plumulaceous, and the alternative locked sheet structure is referred to as being a pennaceous vane. The pennaceous vane structure is what gives feathers their strength. Thus the outer plumage acts as a light but relatively impenetrable armour against wind, water, and abrasion, and the overlapping vanes of the open wing form the solid aerofoil required for flight. The strength of the rachis of tails of tree-climbing species such as the woodpeckers and treecreepers is such that the feathers provide a physical support for the weight of the body of the bird.

One particular class of contour feather, the filoplumes, do have a somewhat atypical structure. They have a rachis that is almost naked, having only a small tuft of plumulaceous barbs at its tip. These feathers protrude through the plumage and are thought to be important in providing the bird with sensory information (via motion-sensing cells at their base) concerning wind movements and feather alignment.

Down feathers and semiplumes

Down feathers and semiplumes (which can be classed as an intermediate between a contour feather and a down feather) do have a calamus, rachis, and vane anatomy similar to that of the contour feathers but their rachis is usually short and the vanes are entirely plumulaceous. As a result they often resemble more of a tuft than a typical feather. These are the feathers that cover nestlings, providing them with excellent insulation, and in some situations with a degree of protection against cannibalism. For example, in some colonial situations, newly hatched (wet) gull chicks are often swallowed whole by neighbouring adults—but when they have dried out their stiff feathers make them difficult to swallow and cannibalism rates decline.

In adult birds, down feathers, and semiplumes are usually found beneath the contour feathers where they continue to perform an insulatory role, and in species of waterbird they contribute towards buoyancy. Interestingly species that experience temperature fluctuations in their annual cycle, such as the Redpoll *Carduelis flammea*, often have a greater number of down feathers immediately after completing their moult in the autumn to provide extra insulation during the colder months; presumably these feathers are lost as a result of wear and tear, or are shed to reduce insulation as the warmer spring and summer proceed.

One particular class of down feather, the powder-down feathers, grow continuously but constantly break at their tip, resulting in the production of the powder of feather wax particles that give them their name. It is presumed that this powder has a role in the maintenance of feather water-proofing.

2.2 Feather tracts

Based upon observations of a living bird about its daily business it could be assumed that the body of a bird is evenly covered with feathers. After all, with

(A)

(B)

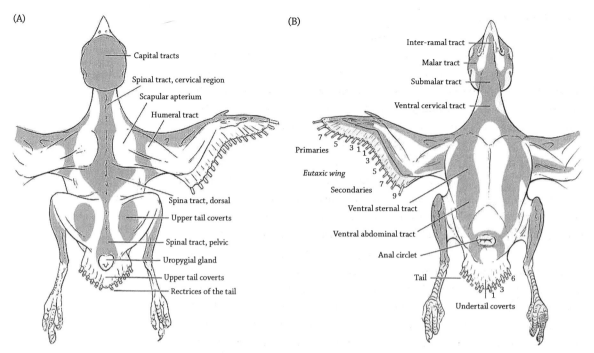

Capital tracts

Spinal tract, cervical region

Scapular apterium

Humeral tract

Spina tract, dorsal

Upper tail coverts

Spinal tract, pelvic

Uropygial gland

Upper tail coverts

Rectrices of the tail

Inter-ramal tract

Malar tract

Submalar tract

Ventral cervical tract

Primaries

Eutaxic wing

Secondaries

Ventral sternal tract

Ventral abdominal tract

Anal circlet

Tail

Undertail coverts

Figure 2.3 The dorsal (A) and ventral (B) feather tracts of a typical passerine bird. Adapted from Proctor N.S. and Lynch, P.J. (1993) *Manual of Ornithology: Avian Structure and Function.* Yale University Press, New Haven.

the exception of the typically bare legs, and the areas adjacent to the beak and eyes, no skin is usually visible. But it would be wrong to assume that this superficial impression of feather-covered skin equates to an even distribution of feathers across the skin in the same way that hair follicles are evenly distributed across a human scalp for example. A similarly even distribution of feather follicles does occur very rarely (examples include the penguins (Spheniscidae) and the ostrich (*Struthio camelus*)) but in almost all birds feather follicles are restricted to well-defined areas of skin—the feather tracts or pterylae which are separated from one another by areas of naked skin, the apteria. Figure 2.3 illustrates the feather tracts of a generalized passerine, and those of a Geenfinch *Carduelis chloris* can be seen in Plate 2.

Along the feather tracts individual feathers grow from specialized groups of skin cells arranged as a feather follicle. These follicles begin as placodes—thickenings of the epidermis and dermis (skin)—which develop to take on a typical 'goose-bump' morphology by a simultaneous evagination of the skin around the feather germ and an upwards growth of the tubular proto-feather as cells at its base proliferate. As it lengthens, layers of cells forming the tubular feather differentiate, the outer cells becoming the protective sheath of the growing feather while the inner cells establish the barb ridges that will form the barbs of the developed feather. Finally the feather bursts from the sheath and unfurls into its final form.

2.3 Feather colour

Birds have a reputation as being amongst the most colourful of vertebrates, and much of this reputation they owe to the incredible range of colours of their feathers. Some colours, white, greens, and blues for example, are the result of the way that structural features of the feather reflect light. For example, the interaction of reflective pigment granules, complex layering patterns in the keratin of the feather, and the angle of the observer relative to the bird are responsible for the iridescence of hummingbirds (Trochillidae) and starlings (Sturnidae).

Other colours, browns, blacks, yellows, and reds for example, are the result of pigments that are laid down within the growing feather. Black, greys, and browns are the result of melanins (specifically eumelanin and pheomelanin)—pigments that are synthesized by birds as a result of the oxidation of the amino acid tyrosine. Darker feathers have more melanin than lighter ones. Most white birds have black, melanin-rich, wing tips; this is because melanin pigments are associated with the deposition of extra keratin which strengthens the feathers (see Box 2.1). Reds, red-browns, and the green colour of turacous (Musophagidae) are derived from porphyrins (specifically turacoveradin (green), uroporphyrin (red), and coproporphyrin III (red-brown)). Porphyrins too are synthesized by birds. In this case they are a product of the breakdown of haemoglobin by the liver. Birds are unable to synthesize the pigments which result in yellows and bright reds (principally leutins and carotenoids); instead they obtain them through their diet directly from the environment. As we will see in Chapter 5 yellows and reds often feature significantly in the courtship plumage of birds, perhaps because they are a signal of male quality.

> **Flight path:** feather colour, male quality, and sexual selection. Chapter 5.

2.4 Feather damage

Although the growing feather does have a blood supply, and the feather itself can be moved by muscles that attach to it below the skin, feathers are inert/dead structures and they cannot be repaired. Feathers may be damaged when birds fight or during encounters with predators or prey; they wear (or abrade) when they rub against one another or against objects in the environment (try pushing your hand through a briar or bramble patch and imagine the abrasions that birds nesting in there must suffer!) (Plate 3). Feathers are degraded as a result of photochemical reactions when the ultraviolet component of sunlight alters the physical structure of the keratin of which they are composed. They are also under attack by an array of bacteria, fungi, and ectoparasites such as mites and lice.

In the face of such relentless pressure why aren't all birds as bald as the proverbial coot! Well of course feathers are absolutely essential to bird survival and so birds engage in regular feather maintenance activities to minimize the impact of damage and wear when it occurs. They also periodically shed and replace all of their feathers in a process known as moult.

Box 2.1   Taking advantage of feather wear

Although feather wear can generally be regarded as a bad thing—necessitating as it does the periodic replacement of feathers—there are birds of a range of species that turn wear to their advantage.

When I catch male Common Redstarts *Phoenicurus phoenicurus* for ringing (banding) in the UK during their southwards autumn migration I am always struck by how dull they are. An adult male redstart during the breeding season is a joy to behold; it is an explosion of colour with its red tail, bright orange/red breast, glossy black face and bib, and powder blue crown and nape (Plate 4A). But in autumn they look, well—dull (Plate 4B). Muted versions of their breeding colours are apparent but overlying them they have a rather dull beige tinge which is often described as a frosting.

I confess that I did initially presume that this was a non-breeding plumage that would be lost during a winter moult in their sub-Saharan African winter grounds. But as I became more experienced (and did some background reading rather than guessing) I discovered that I was wrong. Adult birds of this species complete a full postbreeding moult in the UK before they migrate, and then return the following spring with the same feathers intact. So how do they become brighter? Well if you were able to examine closely the facial feathers of the autumn bird you would see that they are indeed glossy black—but not at their tips—here they each have a pale fringe (hence the beige frosting). These fringes are less durable than the black parts of the feather 'behind' them—possibly because they contain less of the pigment melanin—and they wear away over the course of the winter. Because this wear takes place over all of the body they do in effect become brighter as the breeding season arrives. So come spring they are in prime condition and, without the need to undergo a time- and energy-consuming moult, they are able to get on with the serious business of impressing their mates (and bird watchers like me).

2.5   Feather maintenance

The keratin that feathers are made from is one of the strongest and most durable of biological substances. This is an advantage on one hand, as having resilient feathers is a good thing, but from another perspective can be a problem. When a feather is wrongly aligned, as can happen very easily, it will rub against those around it and might cause increased abrasion. So, one of the most basic of feather maintenance activities undertaken by birds is regular preening (Plate 5). During preening birds smooth their feathers back into place by passing them through their beak. This has the double effect of restoring them to their correct position and 'zipping' back together any barbicles and barbules that have become detached (a process rather like smoothing together the parts of a Velcro fastening). The effect of preening is often enhanced because prior to feather smoothing birds rub onto their beaks the secretions of their preen gland. The preen gland (or uropygial gland) is situated low on the back just above the base of the tail (see Figure 2.3). It produces an oily secretion that is used by birds to maintain the physical quality of their feathers and to regulate bacterial and fungal communities that grow on them. In the case of aquatic birds the preen gland is particularly large and its secretions are important in feather water-proofing. Preening also serves to dislodge and remove dirt particles and ectoparasites, those such as the mites and lice that feed directly on feathers, and also the ticks, fleas, and Hippboscid flies that feed on the birds themselves.

Similarly bathing, in sand or water, scratching (with bill or feet), and having a good shake out can also be useful in feather realignment but are probably most important in the physical removal of parasites and foreign bodies. Sunning (when birds prostrate themselves with their feathers outstretched or stand with their wings open in the full sun) probably also aids in ectoparasite reduction. However, the most intriguing of feather maintenance behaviours are without doubt anting and smoking. In the case of the latter I have often watched with amusement as Jackdaws *Corvus monedula* stand on the lip of a chimney pot and extend first one wing and then another into the rising smoke from the fire below. Presumably the smoke rids the bird of parasites and perhaps an unpalatable smoky residue coats the feathers, deterring the activity of feather-eating mites. During anting, birds position themselves on top of a swarm and allow the insects to crawl through their feathers and over their bodies; presumably the ants pick off parasites. In some cases birds actively select particular ants and rub them onto their feathers; it is assumed that in this case the bird is taking advantage of chemicals produced by the ants that perhaps serve as a parasite deterrent.

2.6 Moult

Moult occurs when a new feather begins to develop in the follicle and simply pushes out the old feather above it. The production of new feathers is expensive in terms of raw materials, and the consequence of lack of resources during feather growth can often be seen in the feathers of passerines that have recently fledged and are still in juvenile plumage. Close examination of the retrices or tail feathers of such birds often reveals fault bars—easily visible lines running across the vanes of all of the feathers and aligned across the width of the tail (Plate 6). These are structural weaknesses caused as a result of a period of resource shortage (perhaps that section of the feather grew on a particularly wet day when the parent bird was unable to provide sufficient food to the growing nestling). The width of the bars is a measure of the duration of the resource shortage and they line up because as a nestling all of the feathers of the tail are grown in at the same time. Possibly because they do grow their plumage in such a short time period the feathers of most juvenile birds are of poorer quality than those of adults, and so a postjuvenile moult involving all or most of the plumage is common. This often involves a change in plumage coloration as birds progress from a more cryptic appearance and/or one that may protect them from the competitive attentions of their elders to the patterns typical of adults of their species. Young European Robins *Erithacus rubecula* do not gain the red breast feathers characteristic of their species until they have dispersed away from their natal territory. This is an advantage because adult male robins are strongly motivated to attack anything that is red in colour.

Fault bars are also occasionally evident in the tail feathers of birds which have undergone a moult into their adult plumage, but these bars will usually only be seen in one or two feathers and they will not line up across the width of the tail. This is because an adult bird moults its tail feathers in sequence, often (but not in all cases) from the centre outwards in pairs. Adult birds almost certainly moult this way both to spread the raw material costs of feather production and to maintain aerodynamic efficiency (see Plate 7). So moult may also be costly because it prevents or impairs normal feather function. A bird with gaps in its tail and wing is less ably to fly efficiently and so may be less able to forage or to avoid predators. This is possibly why moulting birds tend to skulk and be less active, or undergo migrations to specific moulting grounds.

> **Flight path:** flocking is an effective antipredator strategy. Chapter 6.

Extreme examples of this are the annual moults of some species of ducks, swans, and geese, many of which form large flightless flocks on or close to their breeding grounds. These birds are flightless for a period of some weeks between the end of the breeding season and the onset of migration (often adults resume flight to coincide with the onset of flight in their attendant young). By flocking it is likely that these birds reduce the predation risks that they face. In the case of some species of sea-duck such as Steller's Eider *Polysticta stelleri* postbreeding flocks are established in estuaries and on the open sea often with hundreds of thousands of flightless birds congregating to complete their moult.

Box 2.2   Minimizing the impact of moult

The wing of a bird acts as an aerofoil providing both lift and thrust during flight. To maintain flight efficiency it has been assumed that birds optimize their body mass to wing area ratio. A great number of experiments have been carried out to demonstrate that birds can/cannot fly following wing manipulations such as the removal of feathers, feather cutting, and even surgical separation of the propatagium (the flap of skin along the bones of the wing to which the feathers are attached). The results of such mutilations have shown, for example, that pigeons are able to fly in a wind-free laboratory environment with as little as 50% of their wing area intact. But such studies do not tell us much about the real impact of natural wing area reductions, for example during moult. It has been suggested that moulting

birds would suffer reduced flight performance to the extent that they might be less able to forage or to escape from predators. We might therefore have expected natural selection to have resulted in a solution to this problem. Joan Senar and colleagues have carried out a remarkably elegant experiment to show just how Great Tits *Parus major* maintain flight efficiency during a simulated moult, and their experiment was all the more remarkable because they carried it out in the field and because they did not mutilate their birds in any way.

The researchers captured a sample of wild Great Tits and divided them into two groups, A and B. Group A birds had primary feathers 5, 6, and 7 (those in the middle of the outer wing) taped together to simulate an 8% reduction in wing area (a reduction typical during the moult of this species). Group B birds were not taped. All of the birds were weighed and then released. Two weeks later the birds were re-captured and re-weighed. Group A birds then had their tapes removed whilst group B birds were taped (primaries 5, 6, and 7 again), prior to release. After a further 2 weeks the birds were captured for a third time. They were re-weighed, all tapes were removed, and they were all released.

Following the usual strategy for this species as autumn approaches, group B birds increased in mass following their initial capture, but the taped (group A) birds did not (Figure 2.4). After the second capture, group A birds (no longer taped) did increase their mass to that typical of autumn birds, but look at what happened to the newly taped group B birds. Their mass has fallen. These results are evidence that birds strategically alter their body mass in response to a change in wing area to maintain an optimum wing area to mass ratio.

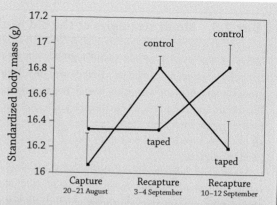

Figure 2.4 Birds with experimentally reduced flight efficiency compensate by adjusting their body mass. From Senar, J.C., Domènch, J., and Uribe, F. (2001) Great Tits (Parus major) reduce body mass in response to wing area reduction: a field experiment. *Behavioural Ecology* **13**, 725–727.

Moult strategies

As has already been mentioned, moult is essential to birds, enabling as it does the replacement of worn and damaged plumage. In some species, however, it serves a further purpose in that it allows birds to change their appearance to suit their needs at a given stage of their life cycle. For example, male birds of a number of species (as diverse as ducks and buntings) annually moult all or some of their feathers to enable them to shift from a non-breeding/eclipse plumage to a breeding/nuptial plumage and back again; and, montane/high latitude birds such as the Ptarmigan *Lagopus mutus* moult into and out of a whiter winter plumage for camouflage.

The particular moult strategy employed by any given species is likely to have evolved as an adaptive response to a range of conflicting pressures such as the availability of resources, the time available before a necessary migration, the need to allocate available resources to reproduction, etc. Some species, particularly those at high latitudes, take advantage of a superabundance of resources and increased day length to breed and moult simultaneously. For example, Ivory Gulls *Pagophila eburna* begin their moult prior to egg laying, and Alaskan populations of Glaucus Gulls *Larus hyperboreus* moult whilst incubating. In the tropics, lengthy periods of parental care, perhaps a result of high competition for patchily distributed resources, often results in the onset of adult moult overlapping the end of

Box 2.3 Moult strategies of the old world warblers

Many of the species of old world warbler migrate annually between northern European breeding grounds and African wintering areas. The timing of the moult and the particular sequence and extent of feather replacement vary from species to species, but all of them have to fit the need to moult around the need to migrate.

Among the *Sylvia* warblers, those species that are relatively sedentary tend to spend longer on their postbreeding moult than do those which undertake long migrations. So, for example, non-migratory populations of the Blackcap *Sylvia atricapilla* take around 80 days to complete the same moult that birds of the migratory population of the UK undertakes in just half of that time. Most *Sylvia* warblers complete their moult on or close to their breeding grounds, but those with longer migrations may start a moult premigration, interrupt it, and then complete it at a suitable staging post *en route*. The birds of those populations which do have a particularly long migration, such as the Garden Warbler *Sylvia borin* which breeds in northern and eastern Europe and winters in Africa some 30° south of the equator, delay their moult until they reach their winter grounds. This is possibly because the distances travelled are so great that Garden Warblers arrive late and are forced to leave their breeding grounds very early in the migration season, and simply do not have the time to complete a moult once they have finished breeding. Once on their winter grounds birds are probably able to complete a moult at a more leisurely pace (not having to fit in a breeding event) and as a consequence are able to begin the northwards return journey with a set of fresh primaries. It has been suggested that this is in itself significant because the spring migration tends to be a bit of a rush to arrive in time to secure the best breeding territories, and in such a race primaries in good condition are likely to confer an advantage.

A similar trend is observed in the moult/migration strategies of the species of another genus of old world warbler, the *Phylloscopus* warblers. The Chiffchaff *Phylloscopus collybita* is a typical shorter distance migrant and one which undergoes a moult before it begins the autumn migration. The closely related Greenish Warbler *Phylloscopus trochiloides*, however, begins its moult before migrating, but completes it once it has reached its winter territory. Notice here that I referred to a winter territory rather than a winter area—this was quite deliberate. Remember that the first birds to arrive secure the best territories. Greenish Warbler establish and defend a winter territory and so presumably are in just as much of a hurry in the autumn as they and the other species I have mentioned are in the spring. Finally another *Phylloscopus*, the Willow Warbler *Phylloscopus trochilus*, moults not once but twice each year. Willow Warblers undergo a rapid and complete moult prior to both the autumn and spring migration. Members of this species undergo particularly long migrations, breeding further north and wintering further south than Chiff-Chaff for example. It is likely that they moult twice because their flight feathers are simply not sufficiently robust to make the trip twice.

References

Shirihai, H., Gargallo, G., and Helbig, A.J. (2001) *Sylvia Warblers*. Helm, London.

Ginn, H.B. and Melville, D.S. (1983) *Moult in Birds*. The British Trust for Ornithology, Tring.

the breeding season. The moults of smaller birds tend to be completed more quickly than those of older birds, and in the cases of some large bird species one moult cycle may overlap with the next and so moult will overlap with the whole of the annual cycle (although it may only be active moult during some periods of the year, being suspended at others).

2.7 Flight

I am confident that there is not a single reader of this book who, when watching birds fly, has not asked the question 'how do they do that?' To remain airborne and fly a bird relies upon two forces—lift and thrust—which must be sufficient to counteract two opposing forces—gravity and drag. Figure 2.6A illustrates the opposing directions of these forces in a hypothetical situation. Gravity acts upon the mass of a bird pulling it downwards towards the earth. To remain aloft, therefore, it must generate sufficient lift. Drag forces (the friction forces which result as the bird moves through the air) will push it backwards. To counter this a bird must produce sufficient thrust either directly through powered flight or more subtly by the manipulation of the lift/drag relationship. Swifts *Apus apus* spend almost their entire life on the wing, coming to earth only to breed. They even sleep on the wing, climbing to high altitudes and flap-gliding through the night. As aerial hunters of insects, speed and agility in the air are essential to them and I have to admit that watching screaming parties of young swifts gathering prior to their autumn migration is one of the thrills of my birding year. During these bouts of aerobatics, and when watching hunting swifts, I am amazed by their ability to gather speed and to perform seemingly impossible direction changes without flapping their wings. They make these manoeuvres by altering their wing shape, that is to say by morphing them. Lentik and coworkers have carried out wind tunnel experiments on the paired wings of dead swifts (the birds died during rehabilitation and were not sacrificed for the experiments). They morphed the wings into a variety of shapes based upon observations of flying birds and found that fully extended wings (long and thin and at 90° to the body) generate maximal lift, but are suited to lower speeds and slow shallow turns. Wings swept back (to 45° relative to the body) generate less lift, but minimize drag and so increase glide speed and enable a bird to turn sharply at speed. So by morphing between these wing shapes (and using intermediates between them) swifts are able to control their glide.

So wings, and specifically the morphology of wings, are the key to flight characteristics, and to achieve the various modes of flight that I am about to discuss there are four basic wing types (see Figure 2.5).

I am not a physicist and do not propose to discuss in detail the aerodynamics of bird wings or the aerodynamic theory of bird flight, but for those readers who do want to explore these areas I can recommend *Avian Flight*, the excellent monograph by John Videler, upon which the following sections of this chapter draw heavily.

As a wing pushes through the air it produces an area of high pressure in front of it. As the air 'splits' and moves quickly backwards across the upper, and less quickly across the lower, surface of the open wing before being deflected downwards, behind it further pressure differentials develop. The net effect is that air

Key reference

Lentik D, Müller, U.K., Stamhuis, E.J. de Kat, R., van Gestel, W., Veldhuis, L.L.M., Henningsson, P., Hedenström, A., Videler, J.J., and van Leeuwen, J.L. (2007) How swifts control their glide performance with morphing wings. *Nature* **446**, 1082–1085.

Key reference

Videler, J. (2005) *Avian Flight*. Oxford University Press, Oxford.

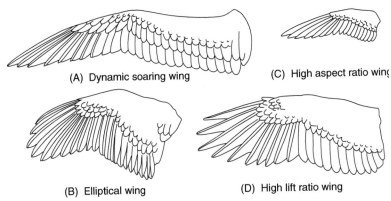

(A) Dynamic soaring wing

(C) High aspect ratio wing

(B) Elliptical wing

(D) High lift ratio wing

Figure 2.5 Dynamic soaring wings (A) are long and narrow, enabling birds such as albatrosses, petrels, and shearwaters to glide at speed. Elliptical wings (B) are broad and rounded; they are typical of birds which require short bursts of speed and high manoeuvrability such as woodland species and those which are the prey of other birds. High aspect ratio wings (C) provide for speed and agility and are commonly found in aerial hunters such as swallows and hawks. High lift ratio wings (D) are broad and fingered; they enable birds such as storks and vultures to soar but have limited manoeuvrability. From Pough, G.H., Janis, C.M., and Heiser, J.B. (2002) *Vertebrate Life*, 6th edn. Prentice Hall, New Jersey. Reprinted by permission of Pearson Education, Inc.

Box 2.4   Adaptations for flight

Birds have a number of key adaptations which enable them to undertake powered flight efficiently. They will be discussed in more detail throughout this book, but the main ones are summarized here:

1. Wings: forelimbs that are modified as aerofoils.
2. A skeleton that is strong and rigid, but that is very light. This is achieved because bird bones are less solid than those of mammals or reptiles. They are filled with a honeycomb of air spaces and strengthened by internal struts.
3. The coracoid bone which acts as a support for the shoulder.
4. An enlarged sternum with a deep keel for the attachment of the large muscles associated with powered flight.

5. Large and powerful flight muscles. The pectoralis and supracoracoideus provide the power for flight.
6. A highly efficient respiratory system involving lungs, air sacs, and haemoglobin with a particularly high affinity for oxygen.
7. Considerable modification of the bones of the forelimb. Fused hand bones, in some cases locking joints and in others wrist joints that rotate almost fully.
8. The furcula or wishbone which acts as a spring during the wing beat cycle.

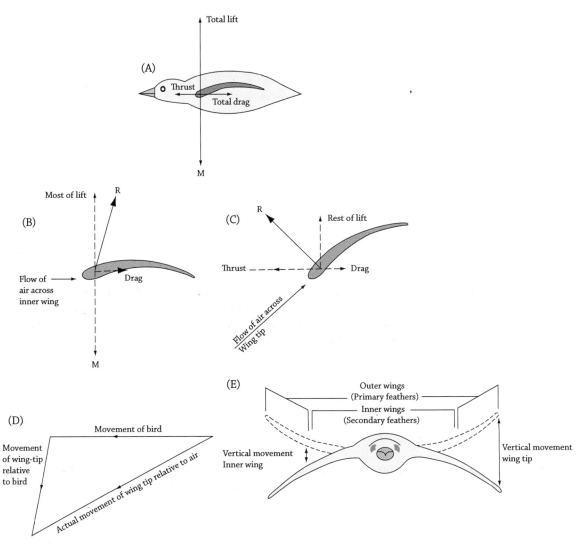

Figure 2.6 The forces acting upon, and generated by, a generalized flying bird. From Pough, G.H., Janis, C.M., and Heiser, J.B. (2002) *Vertebrate Life*, 6th edn. Prentice Hall, New Jersey. Reprinted by permission of Pearson Education, Inc.

moves from the lower pressure areas towards the higher pressure areas, and the forces of lift and thrust result. The balance of these two forces depends upon the angle of the wing relative to the oncoming air stream (Figure 2.6A). When, as in Figure 2.6B, this angle of attack, as it is termed, is low, little thrust is generated but lift is produced. If the leading edge of the wing is tilted forwards (Figure 2.6C and D), both lift and thrust are generated and there is a net movement of the bird forwards. Tilting the wing upwards will of course increase drag and slow the bird. Note that lift is generally produced by the inner wing, as air moves over the surfaces of the secondaries, whereas thrust is produced by the primaries of the outer wing

(Figure 2.6D). The primaries are asymmetrical and each of them acts as an aerofoil in its own right—generating lift additional to that produced by the wing itself. As the wing beats, the primaries twist such that on the downstroke they close to form a solid wing, but on the upstroke they open, reducing wind resistance.

Gliding and soaring

Under the right conditions birds are able to utilize the forces of lift and thrust generated by outstretched wings to travel considerable distances with minimal energy expenditure. They achieve this by gliding or soaring rather than by using energetically expensive flapping flight. Vultures, storks, and other large birds are well known for their ability to achieve altitude by hitching a ride on a rising column of heated air. From the tops of these thermals they are able to soar for long periods to find food or to undertake stages of a migratory flight. By soaring from the top of one thermal to the bottom of another they can travel considerable distances without the need to flap their wings. The most accomplished of the gliders are probably the albatrosses and Giant Petrels of the southern oceans (Plate 10). With their very long, slender wings they too are able to travel large distances, and at considerable speed, with almost no need to flap. To increase energy conservation further they possess a modified wing joint morphology which allows them to lock in position their fully stretched wings. Without this, other species exert muscle energy to achieve the same end. Unlike vultures, who use thermals to achieve lift, these seabirds ride the pressure differentials that result when winds close to the ocean surface move at a slower speed (due to friction) than do the winds above them. This is termed dynamic soaring and relies upon there being a constant wind; a perfect strategy then for the seabirds of the roaring forties.

Smaller birds generally lack the wing area to adopt gliding as a main mode of flight. But they are able to reduce energy expenditure to some degree by punctuating their flapping flight with short glides or bounds. Typically their flight path is sinusoidal, gaining height during a burst of flapping and then falling towards the end of the short glide. But the savings gained by milliseconds of gliding are considerable. As an example, in a study of the bounding flight of the Zebra Finch *Taeniopygia guttata* (Figure 2.7), Tobalske and coworkers have shown that birds were able to reduce their flapping (and presumably therefore their energy expenditure) by between 22% and 45% depending upon their speed of travel.

Flapping flight

The power for flight comes from the flapping of the wings. During a single beat cycle of the wing, in some cases a period of just milliseconds, the wing form, and

Key reference

Tobalske, B.W., Peacock, W.L., and Dial, K.P. (1999) Kinematics of flapping flight in the zebra finch over a wide range of speeds. *Journal of Experimental Biology* **202**, 1725–1739.

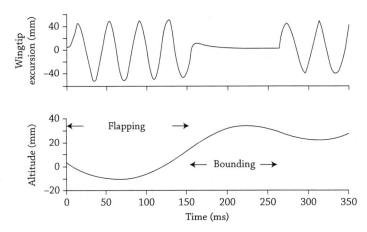

Figure 2.7 The flapping and bounding flight of the Zebra Finch. By flapping, the bird achieves height (altitude) and momentum so that during a non-flapping bound it can conserve energy. Wing tip excursion is a descriptive for flapping activity. Adapted from Videler, J.J. (2005) *Avian Flight*. Oxford University Press, Oxford.

therefore its aerodynamic properties, changes a number of times. The inner wing (the secondaries) simply moves up and down during the cycle and, as it does so, it acts in the same way as a fixed wing during gliding in that it generates most of the lift experienced by the bird. As was mentioned previously, the primary feathers twist during the beat, 'closing' the wing surface on the downstroke and opening it to reduce wind resistance as it moves upwards. At the same time the wrist joint (between the inner wing and the outer wing) twists so that on the downstroke the wing moves both down and forwards and on the upstroke moves up and backwards. So the outer wing is moving through the air in a sort of figure-of-eight motion. As it moves forward and downward on the downstroke, the angle of attack of the wing is high and the lift generated has a forwards direction—thrust.

To hover, to fly at a fixed point, birds have to generate lift sufficient to support their weight but generate no forwards thrust (or at least no more thrust than that required to counter the drag that they are experiencing). For the vast majority this is very difficult and cannot be sustained for anything more than short periods prior to landing, or striking at food. But there is one group, the hummingbirds, that have perfected hovering to the point that they are able to hold their position in mid air for prolonged periods. They can also fly forwards, backwards, and even sideways. This is possible because the wing of a hummingbird has an anatomy unlike that of any other bird (except their close relatives the swifts). The hummingbird inner wing is very short (it bears just six secondaries) and is held in a fixed 'V' position close to the body. The outer wing (with 10 long primaries) forms the main surface and accounts for more than 80% of the wing (compared with about 40% in the Buzzard *Buteo buteo* for example). Their main flight muscles are far larger (relative to body mass) than those of other flying birds. The wrist joint is particularly flexible, enabling the outer wing to twist almost upside down on the back stroke. When hovering, hummingbirds generate lift and thrust on both the forward and back stroke, both of which sweep so far that the wing tips are brought close together. Their wing beat, describing a figure-of-eight in the air that is almost horizontal to the ground, is remarkably fast. Speeds of as much as 200 beats per minute have been recorded from the Ruby-throated Hummingbird *Archilochus colubris* during courtship hovering, although speeds between 10 and 80 beats per second might be more typical of the family.

So we can see the path of the wing during flight and infer from that the movement of joints, etc. But what is going on inside the bird? Figure 2.8 illustrates schematically the main anatomical components of the wing. There are actually 45 different muscles in the bird wing, but only the two thought to be most significant (and currently best understood) are shown. These are the pectoralis and the supracoracoideus, the two relatively large muscles which attach to the deep keel (corina) of the sternum and between them provide the power responsible for the wing strokes of flapping flight. Contraction of the pectoralis pulls the wing down and forwards during the downstroke of flight. The supracoracoideus pulls it upwards and backwards during the upstroke.

The effect of these muscular contractions upon the bones of the wing can be seen in Figure 2.9 which was obtained by Farish Jenkins and colleagues when they trained a European Starling *Sturnus vulgaris* to fly in a wind tunnel whilst being filmed using radiographic film. Essentially they X-rayed it 200 times per second while it flew! From the figure we can see that the humerus is almost parallel with the body at the start of the sequence but that it moves upwards and rotates forwards as the supracoracoideus contracts on the upstroke (Figure 2.9B). The bones of the hand at this point are held at approximately 90° to the body. They remain there as the wing closes (Figure 2.9C) and are swept backwards and down as the contraction of the pectoralis causes the downstroke. This research has also revealed the role of the furcula, or wishbone, during flight. As the humerus rotates forwards and moves downwards, the heads of the furcula move apart. As they bend, the arms of the furcula act as a spring and store some of the force of the downstroke. They then release the stored energy as the furcula returns to its resting position during the upstroke. The exact function of this process is not yet understood. The release of energy may assist the supracoracoideus in raising the wing, but it seems most likely that it has a respiratory function because a relationship between the wing beat cycle, the action of the furcula, and the compression of the air sacs has been identified.

> **Flight path:** energy, flight, and migration. Chapter 3.

Respiration and flight energetics

The lipids (fats), carbohydrates, and proteins of the body provide the energy for metabolic activity, including flight. The 'burning' of these compounds in the

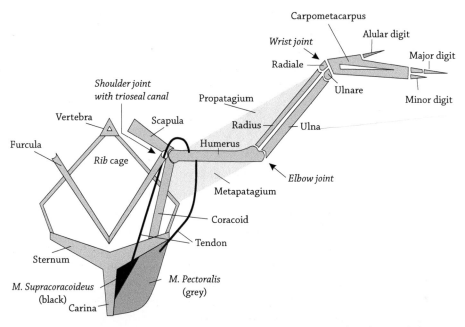

Figure 2.8 An overview of the anatomy of the avian wing and rib cage. From Videler, J.J. (2005) *Avian Flight*. Oxford University Press, Oxford.

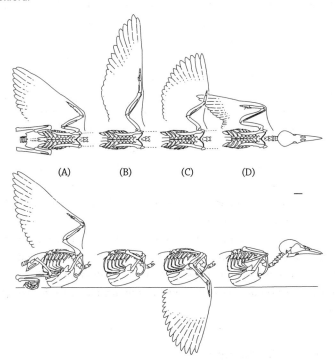

Figure 2.9 Skeletal movements of the European Starling *Sturnus vulgaris* during flapping flight. From Jenkins, F.A., Jr, Dial, K.P., and Goslow, G.E., Jr (1988) A cineradiographic analysis of bird flight: the wishbone in starlings is a spring. *Science* **241**, 1495-1498.

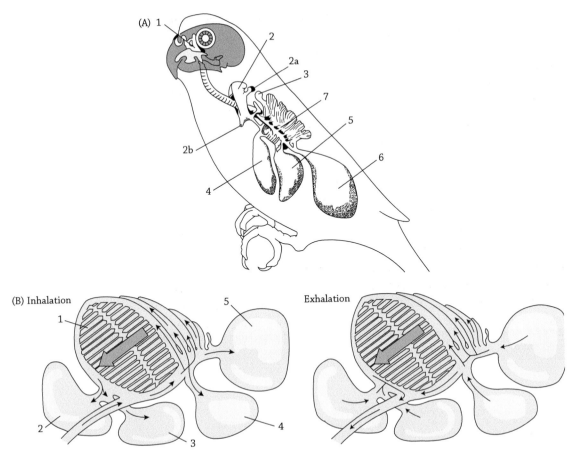

Figure 2.10 Gaseous exchange in the respiring bird. (A) The parabronchial lung (7) and airsac (1–6) system of a generalized bird. Air sacs labelled are: 1, the infraorbital sinus; 2, the clavicular air sac; 3, the cervical air sac; 4, the cranial thoracic air sac; 5, the caudal thoracic air sac; and 6, the abdominal air sacs. (B) The pattern of air flow through the system during both inhalation and exhalation. From Pough, G.H., Janis, C.M., and Heiser, J.B. (2002) *Vertebrate Life*, 6th edn. Prentice Hall, New Jersey.

presence of respiratory oxygen provides the energy required for the splitting of cellular molecules of ATP (adenosine triphosphate) which in turn provide the energy needed to control the contraction of muscle fibres. Oxygen and fuel from feeding (which will be discussed in Chapter 6) are therefore essential for flight. Oxygen is obtained by terrestrial vertebrates when air is inhaled and the oxygen in it is transferred across the walls of the lung to the bloodstream. It is then transported around the body to the cells that need it by the molecule haemoglobin, and the haemoglobin of birds has a particularly high affinity for oxygen. Birds have a higher metabolism than other terrestrial vertebrates, they have higher body temperatures and faster heart rates, and they use more oxygen in flight than mammals do when running. We might therefore expect the lungs of birds to be particularly large, but in fact they are smaller than those of similarly sized mammals. This is probably a necessary adaptation to keep down body mass for efficient flight. So how do birds get the oxygen that they need? Well their lungs may be small—but size for size they have a larger internal surface area for oxygen transport. In mammalian lungs inhaled air passes along bronchioles (tubes) to alveoli (dead-end sacks) where oxygen is absorbed. Exhaled air reverses back along these tubes in what is called a tidal fashion. The lungs of birds are very different. They lack the alveoli and instead the bird lung is composed of a network of very thin tubes termed parabronchi (the bird lung is often therefore referred to as a parabronchial lung), each of which divides into very many thinner capillaries where gaseous exchange takes place. Inhaled and exhaled air passes through the lung in the same direction (i.e. a non-tidal flow) and in fact a complete cycle of respiration involves not one but two breaths. This is possible because the lungs of birds are connected to a network of air sacs within the body cavity (Figure 2.10A).

During the first breath, inhaled air passes into and through the lung and into the abdominal air sac. Contraction of the abdomen during exhalation forces air back out of the abdominal air sacs, back through the parabronchi of the lung, and gas exchange takes place (Figure 2.10B). During the second breath the 'stale' air in the parabronchial lung is forced into the anterior air sacs on inhalation and out of them, and out of the bird, during exhalation (Figure 2.10B). So in a bird oxygen uptake is happening on both inhalation and exhalation. This is a far more efficient system than that of other vertebrate groups and this, coupled with the

---

Box 2.5   Formation flying saves energy

My home is right under the flight path of gulls making their daily journey from our town land-fill site (where they devour our waste) to their coastal and town centre roosts. I rarely take note of single birds, but when they pass in their distinctive V-shaped flocks I regularly have to explain to those around me just why they choose to fly in formation (Plate 8). Quite simply I say large birds fly in a V to conserve energy. I am sure that you have been told the same thing many times, but I guess it might surprise you to know that hard data to support this explanation are actually quite rare. Good data have, however, been recently provided by Henri Weimerskirch and coworkers. They have made detailed observations of the heart rates and wing beat rates of free-flying Great White Pelicans (*Pelecanus onocrolatus*) singly and in flocks. They trained eight birds to fly behind a moving motor-boat and filmed them doing so. The birds had been

fitted with heart rate monitors and from the data from these, and the films that they made, it was possible for the team to compare the individual heart rates and wing beat rates of all of the birds. These measures are presumed to correlate closely with energy expenditure during flight.

Gliding lone pelicans and birds flying 50 m and 1 m above the water flapped more often and had higher heart rates than did birds flying in formation at 1 m (Figure 2.11). Birds at the front of the V have a wing beat frequency similar to that of single birds at the same height, but those behind the leader clearly benefit. However, from the figure you should note that the benefits diminish slightly the further from second place a bird flies. The researchers also noticed that birds at the rear of the formation constantly adjusted their position relative to the group; presumably they were maximizing the energy savings that they made.

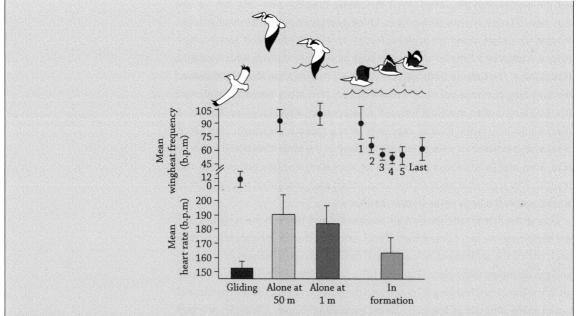

Figure 2.11 Variation in heart beat rate (a measure of energy expenditure) of gliding, solitary, and formation flying pelicans. From Weimerskirch, H., Martin, J., Clerquin, Y., Alexander, P., and Jiraskova, S. (2001) Energy saving in flight formation. *Nature* **413**, 697–698.

increased efficiency of avian haemoglobin as an oxygen carrier compared with that of mammals, might explain why birds are able to migrate apparently effortlessly at altitudes well above the height of Mount Everest, whereas the human climbers on the slopes below them are often reliant upon enhanced oxygen supplies.

Flight speeds

Earlier in this chapter I described the way in which a swift can adjust its flight speed by altering the shape of its wings during a glide. But not all birds are as accomplished at gliding as swifts, and in many cases to fly faster a bird must flap harder and therefore use more energy. This probably isn't a surprise to you. The relationship between power (flapping) and speed seems to be straightforward. It is not, however. Whilst it is true that flying fast is energetically expensive so is flying slowly (remember that hovering is very expensive). Aerodynamic theory suggests that the power curve (the relationship between power and velocity) of flight should be U shaped. By flying Magpie *Pica pica*, Barbary Dove *Streptopelia risoria*, and the Cockatiel *Nymphicus hollandicus* in wind tunnels and directly measuring pectoralis muscle activity, Tobalske, Dial and colleagues have shown that in the case of these species the power

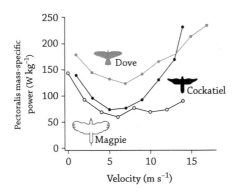

Figure 2.12 U-shaped power curves of three species of flying bird. Note that in all cases slower and faster flight speeds incur a higher energy cost (i.e. more muscle power is required). From Tobalske, B.W., Hedrick, T.L, Dial, K.P., and Biewener, A.A. (2003) Comparative power curves in bird flight. *Nature* **421**, 363–366 and Dial, K.P., Biewener, A.A., Tobalske, B.W., and Warrick, D.R. (1997) Mechanical power output of bird flight. *Nature* **390**, 67–70.

curves which result are broadly U shaped (Figure 2.12). Obviously there are times when a bird will have to fly fast (to catch mobile prey or to escape a predator for example), or perhaps fly slowly (to locate cryptic food or perhaps as part of a display, although in such cases the function of the display may of course be to advertise that you have energy to spare), but based upon these observations we might expect birds to select flying speeds that conserve energy whenever practicable, to adopt optimal flying speeds. In fact it has proven very difficult to demonstrate that birds fly at an optimal speed in real life, perhaps because birds incorporate more information (about their momentum, their motivation, the weather, etc.) into their decision making than we have taken into account when making our prediction.

> **Flight path:** flight can have a display or information exchange function during foraging, territoriality, or courtship. Chapters 5 and 6.

2.8 The evolution of flight and flightlessness

Just as the question of the evolution of birds from dinosaurs has generated controversy (see Chapter 1), so has the question of the evolution of avian flight itself. Several hypotheses have been proposed and abandoned, but two have persisted. One school of thought suggests that flight evolved initially as a means by which a running animal could extend a leap (the ground-up, or cursorial, hypothesis), perhaps to escape a pursuing predator or to capture fleeing prey. But there is limited support for this scenario in either the fossil record or the behaviour of extant flying animals. Currently the hypothesis which seems to be the best explanation is therefore the alternative tree-down or arboreal hypothesis. This hypothesis suggests that flight evolved, initially as gliding flight and then through subsequent modification as powered flight, when animals climbed trees (or cliffs, etc.) and used flight either to slow their fall to

Key reference

Xu, X., Zhou, Z., Wang, X., Kuang, X., Zhang, F., and Du, X. (2003) Four-winged dinosaurs from China. *Nature* **421**, 335–340.

earth or to extend their leap from one high place to another. Evidence in support of the arboreal hypothesis comes from a variety of sources: the fossil remains of many early birds seem better suited to gliding than to powered flight; many of the dinosaur ancestors of birds were able to climb; the evolution of flight in mammals (bats) is thought to have been 'tree-down'; there is a link between climbing and gliding in a wide range of vertebrate taxa; using gravity to provide the initial power flight is easier and more efficient than fighting against it to take off from the ground; the flight stroke needed to prolong a glide is far simpler than that required to lift a bird from the ground; and, perhaps most significantly, recently described fossil dinosaur precursors of modern birds such as *Microraptor gui* had four feathered limbs—exactly what would be expected in an early stage of the evolution of gliding.

It was once presumed that flightless birds had evolved from birds that had themselves never evolved the ability to fly. We know today that this is not the case and that extant flightless forms are in fact derived from flying ancestors. If flight was sufficiently advantageous to the ancestors of modern birds why then should the ability to fly have been lost? Well there are situations where flight is no longer advantageous. The evolution of flightlessness is common, for example amongst those species of terrestrial birds which inhabit isolated islands. In such habitats predators are often absent and so escape flight is not required, nor is it likely that either flight in search of food or flight off the island for whatever reason is particularly important. Amongst marine birds the need to fly may be secondary to the need to swim, and in fact wings may even be a hindrance under water, so they have changed through the course of evolution to become flippers (Plate 9).

Summary

Flightlessness in modern birds has evolved as the loss of flight rather than being the precursor of the ability to fly. To maintain aerodynamic efficiency birds have a number of anatomical and physiological adaptations. Feathers have a range of functions (insulation, display, etc.) but are crucial to flight and have to be constantly maintained and regularly replaced through the process of moult.

Questions for discussion

1. How did avian flight evolve?
2. Discuss a range of moult strategies. How are they incorporated into the annual cycle of the birds exhibiting them?
3. Consider a range of wing forms. What can you deduce from them about the lifestyle/ecology of the species exhibiting them?

CHAPTER 3

Movement: migration and navigation

The stork in the sky knows the time to migrate.

The book of the Prophet Jeremiah,
chapter 8, verse 7

The quotation that opens this chapter demonstrates that the human appreciation of migration is not a new phenomenon. The prophet uses migration as a metaphor to put across a point to his 'audience'. This will only work if that audience understands the substance of that metaphor.

In this chapter I will assume that the reader is more than a little aware of the phenomenon of migration but perhaps less aware of its detail. I want therefore to consider some fundamental points about migration. Why does it happen? How is it controlled? How are the bodies of birds adapted to facilitate migration? What are the consequences of this behaviour both for the birds themselves and in terms of their management and conservation? I also want to consider other movements of birds that whilst not strictly migrations *per se*, do have a lot in common with them. Finally I want to think about the mechanisms that birds use to navigate during migration, and to extend that discussion to consider the navigation of birds during their daily lives.

Chapter overview

If they know nothing else about birds, most people will be able to tell you that some of them migrate. They might not get the detail right, but they will be able to tell you that birds fly south (or north depending on your hemisphere of residence) to avoid bad weather. The general phenomenon of migration, the periodic mass movements of species along established routes, fills us with awe. In fact the level of interest in these long-distance movements is such that in 2004 and 2005 millions of people followed with rapt attention the progress of a handful of Tasmanian Shy Albatrosses *Thalassarche cauta* as they undertook a journey of around 10,000 km across the open waters of the Southern Ocean to and from South Africa. Tracking the birds is possible because they have been fitted with electronic satellite tracking devices and because their daily progress is mapped on an open-access internet site. The project, *The Big Bird Race*, is an innovative collaboration between the business community (in this case the bookmaker Ladbrokes), the Tasmanian State Government, the scientists of The Conservation Foundation, and, in some senses most importantly, by publicity-generating private individuals such as the model and actress Jerry Hall (who sponsored the 2004 winning bird 'Aphrodite') and the publisher Nicholas Coleridge, a direct descendent of Samuel Taylor Coleridge author of '*Rhyme of the Ancient Mariner*' the classic poem in which the fate of an albatross is somewhat prophetically linked to the fate of man.

The bets placed on *The Big Bird Race* raised vital funds towards the 'Save the Albatross Campaign' administered by BirdLife International, and the race itself brought to the attention of the public the plight of albatrosses and of seabirds in general (see Box 3.1). For more information about the race, go to http://www.ladbrokes.com/bigbirdrace.

Not all bird migrations are of the globe-trotting scale of the albatrosses. Another Tasmanian breeding bird, the Swift Parrot *Lathamus discolour*, also undertakes an annual migration. This species breeds in Western Tasmania and feeds largely on the blossom and nectar of the seasonal flowers of *Eucalyptus* species. As the breeding season draws to a close the flowers become less common and the birds range into eastern Tasmania before making a migration northwards across the 300 km Bass Strait into southern Australia. Throughout the southern winter the parrots range across southern Australia in search of food before re-crossing the straits in time to breed the following year.

Concept
Categories of movement

It is possible to recognize two distinct categories of bird movements. The first includes those movements that are concerned with a proximate response to an *actual resource shortage*: foraging trips, commuting, or ranging between patches and dispersal from a natal area to an available local area to establish a home range. These movements conclude when the need for the resource involved is satisfied. The second class of movements are true migrations: triggered by internal rhythms or by a *forecast of resource shortage*. They are characterized by the physiological suppression of the proximate response to resource need and their conclusion is a result of physiological changes resulting from the movement itself.

3.1   The ecology of migration

Why do birds migrate? In the northern hemisphere we tend to think of migrating birds leaving our shores to winter in a place that offers a more benign climate and guaranteed food resources. Similarly we think of the birds that winter with us as doing so to avoid even more harsh conditions at their breeding sites. So we could argue that birds migrate to avoid cold weather. We should remember, however,

Box 3.1   Albatrosses in crisis

The Diomedeidae, the family to which the albatrosses belong has been described as the world's most threatened bird family. The IUCN listed 19 of the 21 species of albatross as being threatened with imminent or close to imminent extinction in 2005. Despite their enigmatic status as lonely wanderers of the oceans, we know surprisingly little about the biology of these long-lived seabirds away from their breeding grounds. One key area of current research involves the use of satellite tracking technology to uncover the migratory and dispersive strategies of individuals and of populations of birds. Members of the British Antarctic Survey have, for example, recently determined that in the case of the adult Gray-headed Albatross *Thalassarche chrystostoma* (Plate 10), three discrete movement strategies seem to be apparent. Some birds stay in their breeding range in the South Atlantic on and around South Georgia. Others make regular return migrations from here to a specific area of the south-west Indian Ocean. Interestingly this area also supports a resident (breeding) population of Gray-headed Albatross, but one that is relatively sedentary. Finally, some South Georgian birds make one or more trips around the world between breeding attempts. Some of these birds fly close to 1,000 km, per day and the fastest recorded circumnavigation took just 46 days (in theory the fastest non-stop circumnavigation would take 30 days). Satellite tracking has also revealed that in addition to annual migrations, individual Wandering Albatross *Diomedea exulans* may make foraging trips of between 3,600 and 15,000 km over up to 33 days when their mates are incubating eggs. Once the chicks have hatched, the trips shorten to around 300 km over 3 days.

It is crucial that we establish the movement patterns of these seabirds in order that conservation strategies might be usefully employed. Unlike many birds, the albatrosses are not endangered because their breeding habitats are threatened, they are dying out because they are the accidental bycatch of human fishing activities. Specifically in excess of 300,000 seabirds fall victim to one fishing technique, longlining, each year. During longlining operations, thousands of baited hooks on a line up to 130 km long are dragged behind a boat. Seabirds in general and albatrosses in particular attempt to take this bait, are hooked, and drown.

The Gray-headed Albatross study suggests that for this species at least only birds staging in the south-west Indian Ocean are likely to come into direct contact with intensive longlining and so perhaps it is in this area that mitigation efforts should be concentrated. The good news is that some mitigation is possible: by setting the lines at night when most seabirds are not foraging or by weighting the lines to sink the bait below the birds' reach, it is possible to minimize the impact that longlining has. Of course this will only happen if boat owners accept and implement these mitigation methods. Towards this end, the governments of 11 countries that practise longlining have already become signatories to *The Agreement for the Conservation of Albatrosses and Petrels* (ACAP) and in doing so have agreed to take specific measures to ensure that their national fishing fleets reduce the impact of longline fisheries and to improve the conservation status of the birds concerned. In light of estimates that between a third and a half of all longlining is being carried out by illegal pirate fishing boats with no specific national allegiance the pressure for change must be maintained lest the albatross become a weight around our collective neck.

Reference
Croxall, J.P., Silk, J.R.D., Phillips, R.A., Afanasyev, V., and Briggs, D.R. (2005) Global circumnavigations: tracking year-round ranges of non-breeding albatrosses. *Science* **307**, 249–250.

that not all migrations are a response to colder winter conditions. In the tropics many species migrate in response to the seasonal patterns of rainfall and drought. But in the case of both tropical and temperate bird species it would appear that migration is at least in part an evolutionary response to predictable climate variability. Migration must therefore also be in part a response to seasonal variations in food resources that accompany these climate fluctuations. But it has also been suggested that migrations may have evolved in response to seasonal changes in the inter- and/or intraspecific dominance relationships between birds—the idea being that at some times of the year the level of competition faced by some birds is such that they migrate to lessen it. You may notice that my language in the preceding paragraph has been quite negative, but perhaps we shouldn't see migration just as a drastic measure taken to avoid hardship. We could equally well think of it as a strategy adopted by birds to allow them to take advantage of a seasonally available opportunity. Think of it this way—should we think of a temperate migrant such as the Sedge Warbler *Acrocephalus schoenobaenus* or the Yellow Warbler *Dendroica petechina* as a temperate bird that tolerates the heat and humidity of the tropics to avoid the northern winter? Or, as a tropical bird that takes advantage of the extended foraging permitted by the longer temperate days?

Whatever the reasons for migration, however, one thing is clear, migration must 'pay' because if it did not birds would not do it. Migration is costly in energy terms and, in terms of risk, many species pass through geographical bottlenecks as they migrate and at these times they are particularly vulnerable to predators including man (only 60% of the wildfowl that migrate south through the USA in autumn each year return to breed the following spring). At least one species of bird of prey, Eleonora's Falcon *Falco eleanorae*, takes advantage of the seasonal glut of migrating prey by timing its breeding to coincide with the autumn passage of passerine migrants out of Europe across the Mediterranean Sea into North Africa. Equally, however, if migration does pay why do some species stay put? Well for each species a balance is probably struck and we can go some way to understanding this balance if we compare key aspects of the life histories of generalized tropical residents, temperate residents, and migrants (Table 3.1).

Key reference

Ketterson, E.D. and Nolan, V., Jr (1983) The evolution of differential bird migration. *Current Ornithology* **1**, 357–402.

Table 3.1 Comparison of key life history traits of generalized migrant and resident species

Trait	Temperate resident	Migrant	Tropical tesident
Productivity	High	Moderate	Low
Adult survival	Low	Moderate	High
Juvenile survival	Low	Moderate	Moderate/high

From Gill F.B. (1994) *Ornithology*. W.H.Freeman & Co., USA.

In their study of the Dark-eyed Junco *Junco hyemalis* Ketterson and Nolan have compared the migration behaviour of males and females and shown that each sex responds in a slightly different way to the selective pressures acting during migration, with the result that this species exhibits a differential migration. Dark-eyed Juncos breed throughout the northern USA and Canada, and populations winter throughout the USA and in southern Canada. But within a population the sexes migrate different distances. Females migrate further south than males. In this case a longer migration is an advantage—females have a higher rate of over-winter survival than males. If migrating a little further south is such an advantage why do males risk dying by travelling less far? Why don't they over-winter in the same place as the females? The answer to this question relates to the fact that for males a second selective pressure exerts a significant influence. Birds that undertake a shorter migration, and survive the winter, have less far to return to their summer breeding grounds. Once there, a strong correlation exists between early arrival on territory and increased productivity. Simply put, the first birds back get the best spots and raise the most young. So there is an advantage to male juncos in not flying too far south that outweighs the risk of dying during the winter.

> **Concept**
> **Migration strategies**
>
> Migrating birds adopt a number of migration strategies. Some species make single long-haul flights, others a series of short hops, refuelling *en route*. Some species are funnelled through migration bottlenecks and others migrate across a broad front. The males and females of some species migrate separately, or migrate to different destinations (often referred to as differential migration). Different breeding populations of a species often use different wintering areas and may leap-frog over one another to reach them.

3.2   Genes and migration

During the migratory period passerine birds exhibit a behavioural change. At night they become restless. This migratory restlessness, or 'zugunrhue', offers researchers a means by which levels of motivation to migrate might be measured and compared.

Captive birds can be housed, singly, in funnel cages such as that shown in Figure 3.1. The floor of the cage is an ink pad and the roof of the cage is a wire screen through which the bird has sight of the night sky. A non-migrating bird will be inactive at night and will stand on the pad. But, during migration, when the bird becomes restless, it will flutter against the sloping sides of the funnel. Evidence of this activity is recorded on the funnel wall by the inky footprints left by the bird. From this relatively simple arrangement two important pieces of information can be gained: the number of footprints equates to the level of restlessness of the bird, or the strength of its drive to migrate; and their position on the funnel wall indicates the direction in which the bird was driven to fly.

As a result of observations made of birds in funnel cages, and of the migratory behaviours of wild birds, it is well established that individuals of a species can differ from one another in terms of their migratory behaviour. From these observations it has been assumed that there is a genetic component to the control of migration. For example. when Schwabl tested in funnel cages the offspring of blackbirds (*Turdus merula*) from German populations known to be migratory or resident, the offspring of migrants demonstrated zugunruhe but those of residents did not. Presumably

> **Key reference**
>
> Schwabl, H. (1983) Ausprägung und bedetung des teilzugverhaltens einer südwestdeutschen population der amsel Turdus merula. *Journal of Ornithology* **124**, 101–106.

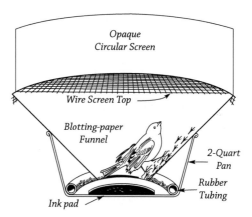

Figure 3.1 A migratory bird held in a funnel cage. The bird's footprints on the side of the funnel indicate its motivation to migrate and preferred direction of flight. From Able, K.P. (2004) Birds on the move: flight and migration. From Berthold, P. (1993). Bird migration, a general survey. Oxford University Press, Oxford.

Key reference

Harris, M.P. (1970) Abnormal migration and hybridisation of Larus argentatus and L. fuscus after inter species fostering experiments. *Ibis* **112**, 488–498.

these birds had inherited their migratory tendencies from their parents. Before we look in more detail at the link between genes and migration it is important, however, to recognize the role that the environment may also play. In Britain the Herring Gull *Larus argentatus* is a sedentary species, while the closely related Lesser Black-backed Gull *Larus fuscus* is migratory. When Harris cross-fostered the young of these species (i.e. placed young *fusucs* in *argentatus* nests and vice versa) he found that the young Lesser Black-backed Gulls raised by Herring Gulls migrated normally—presumably having inherited their migratory behaviour from their genetic parents. However, Herring Gulls raised by Lesser Black-backed Gulls also migrated, although not as far as their foster parents. These birds had not inherited a migratory behaviour pattern and so must have been responding to an external environmental cue—possibly the movement of their foster parents. This demonstrates that whilst genes are clearly important, environmental factors do have a part to play.

The most thoroughly investigated case of the role of genes in bird migration is that of the Blackcap *Sylvia atricapilla*, an old world warbler found throughout Europe and on several North Atlantic islands (the Azores, Maderia, the Canary Islands, and the Cape Verde Islands). Peter Berthold and colleagues have carried out extensive field and laboratory studies of this bird, and no discussion of the genetics of migration could be complete without consideration of their work.

Blackcaps demonstrate a range of migratory behaviours. Populations in northern and eastern Europe are migrants, with most, as we will shortly see, of them spending the northern winter months in Africa. Those of south-west Europe, the Azores, the Canaries, and Madeira are partial migrants (some migrate but others remain on

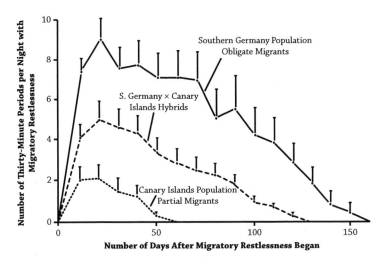

Figure 3.2 Comparison of differing levels of migratory restlessness exhibited by German and Canary Island Blackcaps and exhibited by hybrids between them. From Berthold, P. and Querner, V. (1981) Genetic basis of migratory behaviour in European Warblers. *Science* **212**, 77–79.

their breeding territories all year round). Populations of birds found on the Cape Verde Islands are wholly resident and never migrate.

Berthold and coworkers have compared *zugunrhue* levels of birds from these populations and, as would be expected, have found that birds from Germany (migrants) have higher levels of restlessness than partial migrants from the Canary Islands (Figure 3.2). When hybrids are formed between these two populations the resultant offspring show an intermediate level of restlessness, demonstrating that there is a genetic basis for their migratory behaviour.

Other work carried out by Berthold and his team has demonstrated the genetic basis for migratory orientation—the compass bearing followed by actively migrating birds. Blackcaps breeding in central Europe and migrating to Africa could fly due south and perhaps by doing so minimize their migratory distance. But if they did they would also maximize the time that they spent airborne over the Mediterranean Sea. This sea passage would be arduous indeed and few birds would survive it without significant physiological cost. Instead eastern birds migrate initially in a south-easterly direction, following a route around the eastern edge of the sea and into north-east Africa. Birds from the west on the other hand set off in a south-westerly direction and make the crossing into Africa via the narrow Straits of Gibraltar. Berthold found that hybrids between members of these two populations demonstrate a mixed strategy (Figure 3.3), some flying south-east and some south-west, but some flying due south. From these observations we can deduce that the potential for an initially south-bound migration has been present in Blackcaps from

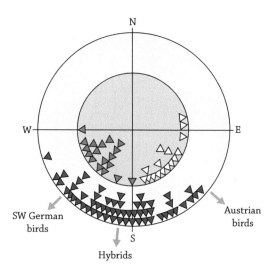

Figure 3.3 Directions of migration of Austrian Blackcaps (inner circle, white triangles), south-west German Blackcaps (inner circle, shaded triangles), and of hybrids between them (outer circle, shaded triangles). From Scott, G.W. (2005) *Essential Ornithology*. Blackwell Science, Oxford, based upon data from Helbig, A.J. (1991) Inheritance of migratory direction in a bird species: a cross breeding experiment with SE- and SW- migrating Blackcaps (*Sylvia atricapilla*). *Behavioural Ecology and Sociobiology* **42**, 9–12.

central Europe, but that it has been strongly selected against, and is not expressed in natural populations.

We are used perhaps to thinking about natural selection as an evolutionary force which has its effect over long, perhaps very long, periods of time. However, as a result of experiments with captive birds it is clear that the results of selective pressures can be seen remarkably quickly. In fact Berthold has shown experimentally that in just three generations of selection for migration a captive population of partially migrant blackcaps can become completely migratory, and in just six generations members of the same population could all be made to behave as residents. Presumably this is exactly what has happened to the resident blackcaps of the Cape Verde Islands, and to the now resident populations of the typically migratory Dark-eyed Junco found on the island of Guadeloupe 250 km off the Californian coast. The loss of migratory behaviour is a common phenomenon amongst species that have colonized remote islands.

The blackcap does offer us one more example of the evolutionary flexibility of migratory behaviour. The British breeding population is a migratory one, but in the closing decades of the twentieth century bird watchers in Britain started to record blackcaps during the winter months and assumed initially that these were British birds that were over-wintering. However, the evidence of ringing recoveries is that these birds are true migrants, members of a population of birds breeding in north-west Europe (Belgium, The Netherlands, and Germany). Why exactly some members of this population should start to migrate north-west rather than south-east in the autumn isn't known, but what is known is that the offspring of these birds share their parents' migratory orientation, demonstrating a genetic component to this novel migratory behaviour.

3.3 Physiology and migration

That there is a genetic component to the control of migration seems certain. But it is important to remember that genes operate in environments, using the word environment here in its traditional ecological sense and also in the sense that the expression of the gene or the activity of its protein product will be affected by the *internal* or physiological environment of the body, and the social environment of the bird concerned.

Control of the onset of migration at this level has been the focus of considerable research in recent decades. But even though advances in our understanding have been made, the detail is not fully understood. It does seem likely that one or more hormones within the bird may have a part to play, but no specific links between particular hormones and specific migratory behaviours have yet been determined. It has been established though that external cues, and in particular seasonal changes in day length, are important. For example, when they compared the rate of development of young Yellow-green Vireos *Vireo flavoviridis* hatched either early or late in the breeding season, John Styrsky and colleagues found that whilst the earlier hatched birds might take as many as 145 days before they began their postjuvenile moult, the latest hatched birds (hatched as many as 7 weeks later) began their moult after just 70 days. This of course results in the birds being relatively synchronous in the timing of their moult (and it turns out in the deposition of fat and the initiation of migratory *zugunruhe*). The birds involved in this experiment had all been collected from the wild when just a few days (6–8) old and then raised in standard conditions. So whatever it was that triggered the accelerated development of the late birds must have had its effect during those first few days of life. It seems likely that the birds even at this early stage were sensitive to photoperiod and took their developmental cue from the day length that they experienced immediately after hatching—a phenomenon described as a calendar effect. Remarkably in this study day lengths across the hatching period varied by just 33 minutes.

We have already seen in Chapter 2 that flight is a metabolically costly activity. Not surprisingly then a key behaviour associated with preparation for and completion of migration is foraging. Birds take on 'fuel' by eating. And prior to a migratory flight birds often engage in bouts of hyperphagia (literally overeating).

Key reference

Styrsky, J., Berthold, P & Robinson, M.D. (2004) Endogenous control of migration and calendar effects in an intratropical migrant, the yellow-green vireo. *Animal Behaviour*, **67**, 1141–1149.

Flight path: Migration and moult are often temporally linked. Chapter 2.

This behaviour tends to take the form of an increase by the individual in the length of time spent feeding each day, rather than individuals eating bigger meals at single sittings (although this may happen too). This increase in the amount of foraging a bird does often coincides with a switch in diet. Birds preferentially select energy-rich foods; many insectivorous songbirds, for example, switch to a fruit-based diet in readiness for migration. In the body the energy from these foods is stored mainly as fat, and so the main fuel for migration could be said to be fat.

Fat is stored in often extensive reserves across the body of a bird, but principally under the skin, in the muscles, and in the parietal cavity. As a fuel, fat is highly effi-cient, yielding almost twice as much energy as protein or carbohydrate, and three times as much metabolic water. On average, birds increase their body fat from around 5% of their total mass when not migrating to 25–35% during migration. These increases in stored fat inevitably lead to increased body mass and, in the extreme case of the Ruby-throated Hummingbird *Archilochus colubris*, a doubling of mass from 3 g to 6 g is needed if this tiny bird is to successfully make an 800 km crossing of the open water of the Gulf of Mexico as part of its annual migratory journey.

There is of course an upper limit to the extra fat that a bird can carry if it is to continue to fly efficiently, and birds have adapted to cope with this limitation in a number of ways. Some species strategically balance their lean muscle to fat ratio to ensure the optimal mass for long-distance flights, and we will look at this in more detail a little later. Other species take advantage of the fact that their migratory routes pass through sites that are ideal as refuelling stations. They store just enough fuel to make the 'hop' from one of these staging posts or stop-overs to the next. In some species the same sites are used by the same individuals year after year.

But not all migratory routes allow refuelling. For example, the geographical distribution of sightings of Bar-tailed Godwits *Limosa lapponica* of the race *baueri* suggested to Robert Gill and colleagues that these birds were likely to make the annual trip from the New Zealand wintering areas to their Alaskan breeding grounds in a series of short hops, refuelling *en route*, but that the return trip was probably completed as a single trans-Pacific flight. Using theoretical models which incorpo-rated information about the known flight metabolism of godwits and the weather systems prevalent along their presumed migration routes Gill and his colleagues proposed that the birds would be capable of making this 11,000 km trip without refuelling. In 2007 their proposal was validated when a satellite-tagged female godwit, E7, made the trip north to Alaska via China in two 'hops', one of 10,200 km and a second of 5,000 km; it then went on to make the 11,500 km return trip in just 8 days (and remember that unlike a seabird a godwit cannot rest at sea).

To achieve these astonishing long-haul trips, the birds metabolize a large fat reserve (migrating birds may carry 41% of their mass as fat), and have evolved a migratory strategy that takes advantage of the fact that the dominant pressure systems across the Pacific during the migration season reliably generate favourable winds. But it is

> **Flight path:** foraging behav-iour is flexible and responds to changing physiological needs. Chapter 6.

Key reference

Gill R.E., Piersma, T., Hufford, G., Servranckx, R., and Riegen, A. (2005) Crossing the ultimate ecological barrier: evidence for an 11000-km-long non-stop flight from Alaska to New Zealand and Eastern Australia by bar-tailed godwits. *The Condor* **107**, 1–20.

Box 3.2 Stable isotopes as a tool to unravel migration

Ringing studies allow us to pin-point exactly where birds go. But they depend upon our ability to capture a bird at one location and then re-capture it at a later date and in another place (perhaps thousands of kilometres away). Re-capture rates are very low for the majority of migratory species—perhaps as few as 1 in 100,000 ringed passerines will be re-trapped in this way. And so the detailed information that we have about the breeding and wintering locations of specific populations of birds, and of the staging posts that they habitually visit during migration, is actually quite limited. Satellite tracking does provide us with more accurate information, but the current costs involved and the limitations of the equipment (size and power principally) mean that even this technology fails to fill as many of the gaps in our knowledge as we might like (but as the technology does advance the possibilities do seem endless).

Chemistry, however, may provide us with an answer to at least some of the questions that we have. You may recall that chemical elements such as carbon (C) each has a specific atomic number, which equates to the number of protons found in the nucleus of one atom of that element—so carbon, for example, has 12 protons and the atomic number 12. But whilst the number of protons in a carbon atom is constant, the number of neutrons is not. So carbon can exist in various forms or isotopes. Examples are carbon-12 (^{12}C), carbon-13 (^{13}C), and carbon-14 (^{14}C). Atoms of these isotopes all have 12 protons, but a ^{12}C atom has 12 neutrons, ^{13}C has 13, and ^{14}C has 14. During photosynthesis plants fix atmospheric carbon, and the ratio of ^{12}C to ^{13}C that is fixed in this way can be characteristic of particular vegetation communities. This particular ratio can be detected in the carbon-containing tissues of birds that have grown while they were feeding on plants, or on animals that had themselves fed on plants, in an area dominated by that vegetation community type. So if we trap a bird in one area of known ^{12}C:^{13}C ratio and find that its feathers exhibit a different ratio we can be sure that it grew them elsewhere. In fact the distributions of these plants results in the existence of a latitudinal

gradient of the isotope ^{13}C. A latitudinal gradient of the hydrogen isotope deuterium also exists (related to a latitudinal gradient in annual rainfall), and gradients or variations in the isotopes of other elements also exist. By comparing information from the isotopes of a range of elements in bird tissues with those from the environments through which a bird might have travelled, we might be able to pin-point the areas that a bird has actually visited more precisely.

But does the theory work? Well the answer to this question is yes it would appear that it does. As an example, consider the work of Chamberlain and coworkers (see Figure 3.4). They have collected feather samples from Willow Warblers *Phylloscopus trochilus* (Plate 11) of two recently diverged (in evolutionary terms), but discrete sub-species *P. trochilus trochilus* and *P. trochilus acredula*. Willow Warblers are common throughout Europe, but the birds that they studied come from an interesting Swedish population where the ranges of the two sub-species come into contact. *acredula* breed in the north of the country, *trochilus* breed in the south, and the breeding ranges of the birds overlap slightly at 62°N latitude. Limited evidence from ringing recoveries suggests that the populations of the two sub-pecies winter in different areas of sub-Saharan Africa (see Figure 3.4). If this is the case, one might expect tissues produced during the winter to vary in terms of their isotope ratios in a way that would reflect variation in the environmental isotope ratios in these different regions. Willow Warblers undergo a complete moult during the winter and when feathers from birds that had returned to Sweden after migration were analysed it was found that the make-up of the feathers of the two different sub-species varied significantly in terms of their carbon and nitrogen isotopes. What is more, the researchers found a sharp change in their data in samples collected on ether side of the 62°N line of demarcation between the breeding populations. So the evidence of the chemistry supports the evidence of ringing studies and the technique could be important in making the links between far distant habitats that are essential to specific populations of migrating birds.

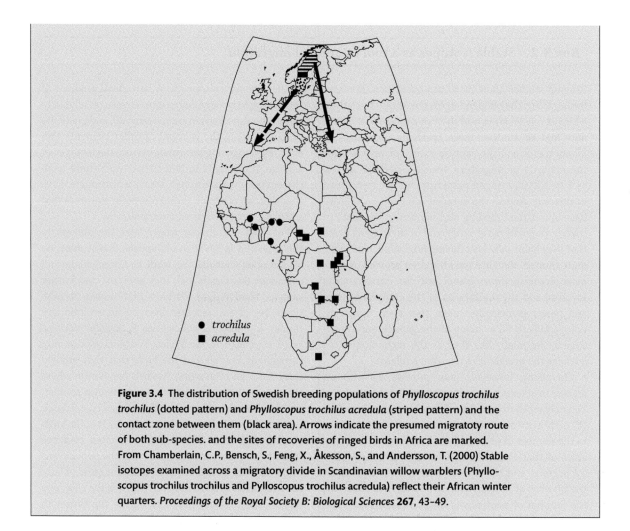

Figure 3.4 The distribution of Swedish breeding populations of *Phylloscopus trochilus trochilus* (dotted pattern) and *Phylloscopus trochilus acredula* (striped pattern) and the contact zone between them (black area). Arrows indicate the presumed migratoty route of both sub-species. and the sites of recoveries of ringed birds in Africa are marked. From Chamberlain, C.P., Bensch, S., Feng, X., Åkesson, S., and Andersson, T. (2000) Stable isotopes examined across a migratory divide in Scandinavian willow warblers (Phylloscopus trochilus trochilus and Pylloscopus trochilus acredula) reflect their African winter quarters. *Proceedings of the Royal Society B: Biological Sciences* **267**, 43–49.

Key reference

Landys-Ciannelli, M.M., Piersma, T., and Jukeman, J. (2003) Strategic size changes of internal organs and muscle tissue in the bar-tailed godwit during fat storage on a spring stopover site. *Functional Ecology* **17**, 151–159.

likely that as in the case of other godwits and other species of bird, some protein metabolism is also necessary—with the effect that body condition may suffer.

Bar-tailed Godwits of the race *Limosa lapponica tamyrensis* may not undertake as impressive a long-haul migration as their *baueri* 'cousins', but they do make a 9,000 km migration between breeding grounds on the Russian Taymyr peninsular and the mud flats of west Africa, and in particular those of Mauritania and Guinea Bissau. However, these birds do not make the trip as a single flight. Instead they divide the journey into two approximately equal flights each lasting around 60 hours. Between these two flights they spend a month refuelling on the rich mud flats of the Dutch Wadden Sea.

When Landys-Ciannelli and coworkers examined the bodies of Bar-tailed Godwits at various points during the refuelling period they found that the amount of fat birds carried increased with time (from around 10% to 30% of total body mass). And that is exactly what one would expect if during refuelling birds are replacing fat used up on the first leg of the journey and then laying down a store for use on the second. They also found some extremely interesting changes in muscle mass that suggest that the birds may be varying the mass that they carry in a strategic way. The mass of the muscles associated with flight varied in line with fat load. Birds had a lower muscle mass when they arrived at the stop-over site than they did when they left—so we can assume that some flight muscle is lost during flight and presumably birds build up their flight machinery so as to be as prepared as possible for the coming journey. Variations were also found in the mass of the stomach, kidneys, liver, and intestines, i.e. in the organs associated with digestion. Newly arrived birds arrived with low mass and very quickly gained mass in all of these components of the digestive system. We would expect them to do so given that they are about to engage in a bout of prolonged hyperphagia. But unlike the continuing weight gain observed in fat stores and in flight muscles, this weight gain peaked during the early part of the refuelling period—mass remained constant during the middle period, and then as departure neared mass fell again as the digestive system atrophied. In this way the birds seem to manage their 'baggage allowance'—they do not carry extra mass associated with a redundant function (godwits cannot feed on the wing) and must therefore 'save fuel'.

> **Flight path:** birds strategically manage their mass to maximize flying efficiency. Chapter 2.

3.4   The weather and migration

Observations of flocks of duck during the migration season seem to suggest that birds wait for good weather before setting off on migratory flights. During periods of poor visibility, heavy cloud cover, and unfavourable head-winds they stay grounded. Then, when conditions improve so that visibility improves, cloud thins, and a tail-wind builds they take to the air. So good weather is important in the initiation of migration, and it remains important as migration proceeds. We have already seen that the extreme migration of some populations of Bar-tailed Godwit is made possible at least in part by the fact that their crossing of the Pacific coincides with reliable tail-winds. Similarly it has been suggested that small passerines are only able to cross the Sahara during their autumnal south-bound migration by flying with the wind.

Of course our own experience tells us that although we might be able to generalize that the winds in a particular season are generally from a certain direction, the day to day specifics of weather conditions are far more variable. As a result, it is not uncommon for individual birds and flocks of birds to be pushed off course or grounded with sometimes disastrous results when the weather becomes unfavourable. In his excellent book *Weather and Bird Migration* Norman Elkins describes an extreme fall (fall being the term used to describe a mass grounding of migrating birds) that occurred in south-east England on a single day in 1965. High pressure over Scandinavia had created the ideal conditions for the initiation of migration, but as they crossed the North Sea the migrating birds encountered a warm front associated with a small but intense low pressure system. Encountering cloud, rain, and unfavourable wind the birds were forced away from their usual southwards route. Thousands of dead birds were washed up along the English coast; presumably their energy reserves had run out as they struggled against the wind. But enormous numbers of birds did make landfall—one recorder estimated more than 30,000 displaced migrants along just 4 km of the coastline and half a million along a 40 km stretch. Reports of the day even suggest that such was the shortage of trees in some towns that birds fluttered onto the shoulders of people!

> **Key reference**
>
> Elkins, N. (1983) Weather and Bird Behaviour. T&D Poyser, Calton, UK.

3.5 Navigation

We tend to think of migrants as having special status in terms of their navigational abilities. There is something awe inspiring about the fact that an 8 cm long passerine can return with pin-point accuracy to its breeding area having first flown across the best part of two continents to get there. But surely the ability of a bird to relocate its nest again and again during the course of the foraging day or to relocate a cached seed accurately in a complex habitat is equally amazing? At both scales these movements require the individual to navigate.

At the local scale there is evidence that birds make use of visual landmarks, a class of navigation referred to as '*piloting*'. The evidence for this comes from experiments in which homing pigeons are fitted with frosted contact lenses prior to being released some distance from their home loft. These birds, by the means that we will discuss below, are able to return to the general vicinity of the loft. However, being unable to see it, they simply land and wait to be carried in.

> **Key reference**
>
> Perdeck, A.C. (1958) Two types of orientation in migrating Starlings Sturnus vulgaris L. and Chaffinches Fringilla coelebs L., as revealed by displacement experiments. Ardea **46**, 1–37.

Navigation at the larger scale, involving the types of orientations and movements characteristic of migrations, seems likely to involve two processes: a compass orientation—often termed '*vector navigation*'—and a goal orientation which is often termed '*true navigation*'. The distinction between these is possibly best explained by reference to an example, and in this case no discussion of the topic could be complete without reference to a classic example afforded by the work of Perdeck in the 1950s.

Perdeck captured more than 10,000 starlings *Sturnus vulgaris*, a mixture of adults and juveniles, in The Netherlands at the onset of the autumn migration. Typically these birds would have undertaken a relatively short south-westerly migration to winter in the southern half of the UK and along the northern coasts of Belgium and France. The birds were marked so that their movements could be tracked, and then transported almost 400 miles to the south-east prior to release in Switzerland. During the following winter around one-third of these birds were relocated, having completed their migration. But where had they gone? In fact, as can be seen from Figure 3.5, two types of movement had taken place. Some of the birds had flown in a north-westerly direction towards their traditional wintering grounds and a few had even made it there. Others had found novel wintering grounds in southern France and on the Iberian peninsular. The birds had separated into two discrete populations according to age; the juveniles had headed south-west while the adults had flown north-west. Remember that juveniles would not have previously migrated and had no experience upon which to draw. Remember also that earlier in this chapter we saw that the initial migratory orientation is innate. So it would seem that these birds were able to set off in the right direction using a compass sense, but that lacking a map sense they had no way of knowing that they were going to the wrong place. This is an example of vector-based navigation.

The adult birds, on the other hand, had presumably migrated to and from the traditional wintering area at least once before, and in doing so had established a map sense which allowed them in some way to establish a link between their current location and their ultimate goal. Translocated to Switzerland, they demonstrated true navigation and were able to undertake a journey across unfamiliar country to reach a familiar goal.

It seems likely that during their first migration young birds learn their route and the locations of their breeding and non-breeding areas with reference to cues from the environments around them. With this acquired information they are then able to develop their mental map and then to use it during subsequent journeys. Several cues appear to be important, and it seems likely that when they are available they are used in combination.

Navigational cues

Birds navigate by day and by night, and use celestial cues to enable them to do so. By day the sun is the dominant cue and at night it is replaced by the stars. That the sun can be used as a navigational cue was first demonstrated by Kramer in the 1950s. In a series of influential experiments he first established the preferred migratory orientation of *zugunruhe*-exhibiting starlings (through the use of funnel cages). He then positioned mirrors around the cages in order to shift (from the point of view of the birds) the position of the sun by 90° and found that the orientation of the birds changed to compensate for this. Further experiments have established that the specific cue used is in fact the position of the sun relative to the horizon, or more precisely the position of an imaginary line from the sun to the horizon referred to as the azimuth.

Of course the sun is not a stationary body. The azimuth moves during the course of the day and so birds using it for navigation must use a bi-coordinate system that takes account of both solar position and time. Clock-shifted starlings, i.e. birds that have been conditioned in captivity to be out of sync' with natural daytime, have

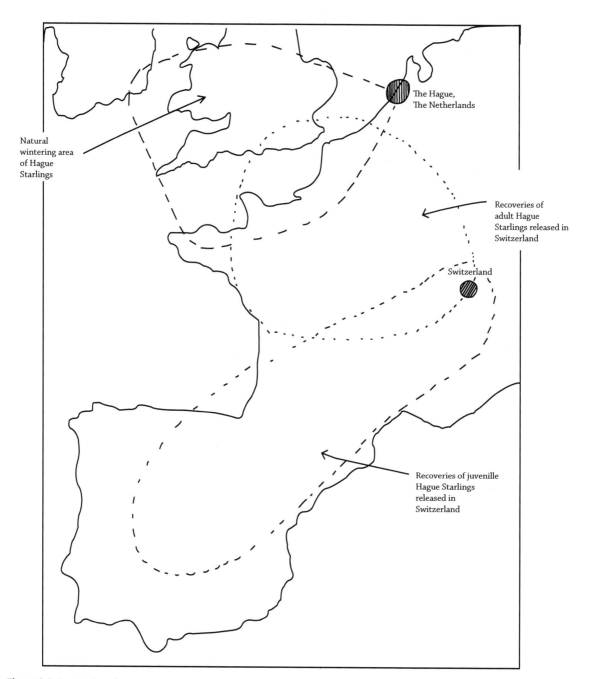

Natural
wintering area
of Hague
Starlings

The Hague,
The Netherlands

Recoveries of
adult Hague
Starlings released in
Switzerland

Switzerland

Recoveries of juvenille
Hague Starlings
released in
Switzerland

Figure 3.5 Recoveries of migrating juvenile and adult European Starlings *Sturnus vulgaris* following their translocation from The Hague (The Netherlands) to Switzerland. Adult birds compensated for their new start point, while juveniles did not. Re-drawn from data in Perdeck, A.C., 1958. Two types of orientation in migrating Starlings *Sturnus vulgaris* L. and Chaffinches *Fringilla coelebs* L., as revealed by displacement experiments. *Ardea* **46**, pp 1-37.

Plate 1 The tail feathers of this male Indian Peafowl are used as an impressive courtship display. © Ingram.

Plate 2 This adult European Greenfinch *Carduelis chloris* is moulting. Lines of new contour feathers just emerging from their sheaths clearly show the position of the ventral sternal feather tract. © Peter Dunn.

Plate 3 The tail feathers of this Whitethroat *Sylvia communis* exhibit extreme wear. This species skulks in dense and often thorny vegetation and its feathers become very worn prior to its annual moult. © Peter Dunn.

Plate 4 In spring male redstarts have a glossy black throat, grey crown, and rusty red breast (A; © Ian Grier). This bird was trapped in Cyprus in spring during its northwards migration. On the other hand the bird shown in B (© Peter Dunn), trapped in the UK during its southwards autumn migration, exhibits the dull fringing typical of freshly moulted birds.

Plate 5 Roosting birds such as this Dunlin spend a considerable proportion of their time carefully preening their feathers. © Ian Grier.

Fault bars

Plate 6 A series of faults are evident in the tail feathers of this Chaffinch *Fringilla coelebs*. Because they line up across the tail we can deduce that this is a juvenile bird. The lack of bars on the three outer feathers of the right half of the tail suggest that these feathers have been replaced. © Peter Dunn.

Plate 7 An example of wing moult in the European Starling *Sturnus vulgaris*. This bird was trapped for ringing in the late summer and is moulting out of its brown juvenile plumage into its glossy adult plumage. Using the numbering conventions explained in Figure 2.1, the wing feathers are dropped and replaced sequentially starting with primary 1 (in the centre of the wing). In this case three new (darker) feathers are visible (primaries 1, 2, and 3), one feather (primary 4) is missing and primaries 5–9 (older brown juvenile feathers) can clearly be seen. The small 10th primary is not visible. Secondary moult has not yet started. © Graham Scott.

Plate 8 Formation flying increases individual flight efficiency. © Ingram.

Plate 9 Using their modified wings as flippers, penguins 'fly' through the sea. © Corbis/Digital Stock.

Plate 10 Gray-headed Albatross. © Ian Robinson.

Plate 11 Willow Warbler *Phylloscopus trochilus trochilus*. © Ian Robinson.

Plate 12 Although obvious when pointed out, this oystercatcher nest was particularly difficult to locate on the shore. © Graham Scott.

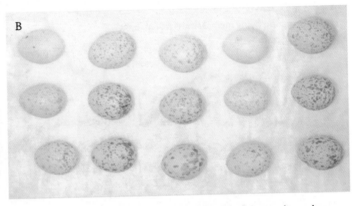

Plate 13 A Great Tit *Parus major* clutch (A) and eggs from five clutches to illustrate their variability (B). The eggs shown here in 'columns' are from the five separate clutches, and the 'rows' are the first (top), middle, and last (bottom) eggs of the laying sequence in each of them. © Andrew Gosler.

Plate 14 The mud nest of the Red-rumped Swallow *Hirundo daurica*. © Robin Arundale.

Plate 15 The intricately woven nest of a weaver bird (Ploceidae). © Graham Scott.

Plate 16 This Oystercatcher *Haematopus ostralegus* has just adjusted the position of her eggs and is settling down to continue their incubation. © Graham Scott.

Plate 17 The unfeathered belly skin of the European Goldfinch *Carduelis carduelis* is stretched taught and heavily vascularized to create a brood patch. © Chris Redfern.

Plate 18 (A) These newly hatched semiprecocial Arctic Tern *Sterna paradisaea* chicks will stay in their nest for only a few days. © Ian Grier. (B) In contrast, these altricial Cormorant *Phalacrocorax carbo* chicks will be nest bound for several weeks. © Les Hatton and Shirley Millar.

Plate 19 Presumably this male Common Tern *Sterna hirundo* provided a sufficient gift and impressed its mate. © Peter Dunn.

Plate 20 Great Grey Shrike *Lanius excubitor*. © Ian Robinson.

Plate 21 (A) Male Red-winged Blackbird *Aegaius phoecniceus*. © Ian Robinson. (B) This Red-winged Blackbird nest was built on a platform designed by Searcy and Pribil to provide females with high quality overwater nest sites. © Stanislav Pribil.

Plate 22 This male Common Whitethroat *Sylvia communis* is in full song, proclaiming ownership of a territory and availability to mate. © Ian Robinson.

Plate 23 The bright yellow gapes of these begging Barn Swallow *Hirundo rustica* are an effective signal of hunger to their hardworking parents. © William Scott.

Plate 24 Having caught a fish, osprey carry them to a safe perch at which to eat them. © Photodisc.

Plate 25 A hovering Malachite Sunbird taking nectar from *Nicotina*. © Sjirk Geets.

Plate 26 When immobile this Ringed Plover chick is almost indistinguishable from the pebbles around it. © Graham Scott.

Plate 27 This bundle of feathers is a perfectly fit adult Ringed Plover feigning extreme injury. Seconds later it took flight. © Graham Scott.

Plate 28 The Northern Pygmy Owl *Glaucidium gnoma* is a significant predator of chickadees and other small birds. © Photodisc.

Plate 29 Through citizen science programmes such as bird counts even the youngest of ornithologists can make a positive contribution to bird conservation. © Graham Scott.

been used in experiments to demonstrate this. If starlings are conditioned so as to be clock shifted by 6 hours (so that at noon they 'think' that it is in fact 6 am) they will demonstrate a 90° shift in their orientation. By this I mean that when tested against the natural position of the sun at 6 am they will fly in an inappropriate direction for the real time of day. Because the azimuth moves by 15° per hour, they demonstrate the 90° shift and fly in a direction more appropriate to a flight under-taken at noon. So it would appear that these birds do cross-reference the position of the sun and time when direction finding.

At night the sun is no longer available and in its place the stars are used instead. Experiments involving funnel cages in a planetarium (so that the star map experienced by the birds can be manipulated) have shown that young birds observe the apparent rotation of the pattern of stars above them as the earth rotates beneath them. In this way they learn the position of the centre of the map, which in the northern hemisphere is Polaris, the North Star. This means that once they have learned this they are able to determine north correctly as long as they can see that star and some of those around it. With this information northern hemisphere migrants can fly south (away from the star) in the autumn and north (towards it) in the spring. But what happens when clouds obscure their view? There is some evidence that suggests that birds may choose not to migrate on cloudy nights. Blackcaps *Sylvia atricapilla* and Red-backed Shrikes *Lanius collurio* have both been observed to show zugunruhe on cloudless nights but not under an overcast sky. However, there is also considerable evidence that birds make use of a range of other cues to compensate for the 'loss' of the sun or the stars because birds are able to orientate appropriately in the absence of a visible sky.

Homing pigeons have no problems finding their lofts on sunny days when they are made to fly with a bar magnet attached to them, but they lose the ability to home when carrying the magnet on a cloudy day. The magnet will cause a localized disruption of the magnetic field experienced by the bird and so it seems likely that geomagnetism is important in orientation (presumably the pigeons are unaffected on sunny days because they are able to utilize their sun compass).

The magnetic poles of the earth maintain a predictable magnetic field across the surface of the planet with a south to north orientation. At the poles the field dips towards the surface of the earth at an angle of 90°, and at the equator the angle is 0°. Birds have been shown experimentally to be able to detect this field, or more precisely to detect the angle of dip of the field as it varies with latitude. This should in theory allow them to establish the directions north and south, and to position themselves on the surface of the earth. For example, European Robins *Erithacus rubecula* have been shown to reverse their orientation when exposed to an artificial reversal of the magnetic field, and Wolfgang Wiltschko has shown that a strong magnetic pulse is enough to disrupt the migratory orientation of Silvereyes *Zosterops l. lateralis*. As Figure 3.6 shows, these birds usually follow a

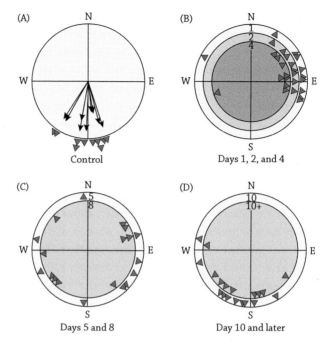

Figure 3.6 Migratory direction preferences of silvereyes prior to (control) and following exposure to a strong magnetic pulse. It is clear that over time the disruptive effect of the pulse wears off. From Scott, G.W. (2005) *Essential Animal Behavior*. Blackwell Scientific, Oxford.

southerly route from their non-breeding range on the Australian mainland to their Tasmanian breeding grounds. A magnetic pulse disorientates them for some days, but the data do suggest that they are able either to recover from the disruption, or to at least accommodate its effect after 10 days or so. Such a disruption may be possible under natural conditions as a result of a solar storm for example.

Exactly how birds are able to detect magnetic fields remains a matter for speculation. A specialist magneto-receptor organ has not been found, but magnetite crystals (Fe_3O_4) in some cells do align themselves according to a magnetic axis and are a clear candidate for ongoing investigation. It has also been shown that the sites of activity in the brain when birds do respond to changes in magnetic fields are in the visual system and that they can only be recorded if the retina is intact and operational. So pigment cells and the photopigments that they contain are also under active investigation.

3.6 Spatial memory

Although efforts to determine the exact nature of the 'magneto-receptor' have yet to come to fruition, attempts to elucidate the means by which birds store and use spatial information have had more success. Experiments involving a range of vertebrates have shown that specific areas of cells in one region of the brain, the hippocampus, are essential to tasks involving spatial memory. In rats, a class of hippocampal cells termed '*place cells*' have been shown to become active in response to the animals encountering a specific landmark, or to groups of landmarks that have a specific relationship to one another and to the spatial position of the test animal. In pigeons, hippocampal lesions (which destroy these cells) disrupt the ability of the birds to home even short distances over familiar ground. Such birds do set off in the right direction so the hippocampus does not appear to be related to compass sense, but they get lost *en route*, presumably having lost their map sense. If the lesioned birds are confined to their lofts for a sufficient period they will eventually regain their ability to home to it, but they will never learn to home to a new loft. The hippocampal cells are clearly therefore involved in both the acquisition and storage/retrieval of spatial information.

Box 3.3   Finding stored food

Important evidence for the key role of the hippocampus in spatial memory comes from comparative studies of the foraging behaviour of passerine birds. The members of some passerine species are well known for their ability to hide and then retrieve seeds. This caching behaviour is obviously to the advantage of the birds concerned. Caching allows them to take advantage of a greater proportion of a food resource than might be possible without storage. It may also provide them with the food they need to live through the lean times, but only of course if they can remember where they hid the food.

David Sherry has shown experimentally that captive Black-capped Chickadees *Parus atricapillus* can do just this. He provided birds with an opportunity to store sunflower seeds in 70 holes drilled into posts in an aviary. When the birds had cached four or five seeds they were ushered out of the aviary and Sherry cleaned it, removed the seeds and then covered every one of the holes with an identical Velcro flap before allowing the birds to return. In the wild chickadees are constantly on the move, lifting bark, turning leaves, and probing crevices—opening Velcro flaps would be second nature to them. So when they got hungry would the birds remember where the seeds had been

hidden and open the right stores? Yes they did. They spent almost ten times as much time exploring the holes in which they had stored seeds compared with the sites they had not used. They were also far more likely to visit a previous cache than an unused site. Taken together, these observations strongly suggest that the birds do remember where they have hidden food. Further work, in the field, has shown that the birds only use each cache once, put just one item of food into each cache, and can remember where they have hidden a seed for almost a month.

The role of the hippocampus in this context has been established via a range of studies, but I think that one in particular is worthy of some consideration, providing as it does an excellent example of the *comparative approach* in biology. Sue Healy and coworkers have compared the foraging behaviour and related brain structure of two species of European Parid closely related to the Black-capped Chickadee. Their work involved the Blue Tit *Cyanistes caeruleus* and the Marsh Tit *Parus palustris*. Marsh Tits, like the chickadee, are avid food storers. Individual birds are known to store up to 100 seeds in a morning, and across the course of a typical winter will store, and perhaps more importantly retrieve, literally thousands of items of food. Blue Tits, on the other hand, do not store food and given this fundamental difference in the foraging ecologies of the two species we might hypothesize that they will have different spatial memory abilities and adaptations. Specifically Healy and her coworkers have asked the question do they have differently developed hippocampi?

In an attempt to answer this question the researchers have compared the hippocampal volumes of birds of both species. Consider first the data in Figure 3.7 which relate to juvenile birds of both species. The data show that when hippocampus size is expressed relative to the size of the telencephalon (an area of the brain not related to spatial memory and therefore not expected to vary between these species) bigger brained juvenile Marsh Tits do have a bigger hippocampus, but so do bigger brained juvenile Blue Tits, and size for size there is no difference between them. This might seem to be a disappointing result, but look at the figure again and this time pay particular attention to the data collected from adult birds. The hippocampus volume of the adult Blue Tits is no different from that of either juvenile Blue Tits or juvenile Marsh Tits, but that of the adult Marsh Tits is appreciably larger than all of these. So why is there a difference in adults but not juveniles? It seems that the juvenile Marsh Tits had not yet had the opportunity to store food whereas the adult birds had. In effect the juvenile Marsh Tits were behaving ecologically more like a typical Blue Tit. It

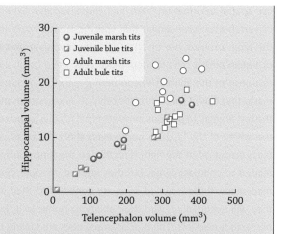

Figure 3.7 Adult Marsh Tits who have had experience of food storing have a more developed hippocampus than do inexperienced juvenile Marsh Tits or adult/juvenile non-storing Blue Tits. From Healey, S.D., Clayton, N.S., and Krebs, J.R. (1994) Development of hippocampal specialisation in two species of tit (Parus spp). *Behavioural Brain Research* **81**, 23–28.

would appear, therefore, that a larger hippocampus is related to spatial memory, but that hippocampal enlargement is a response to food-storing behaviour rather than a prerequisite for it.

Key reference

Sherry, D.F. (1984) Food storage by black-capped chickadees: memory of the location and contents of caches. *Animal Behaviour* **32**, 451–464.

Summary

Migration allows birds to make the most of the resources available to them. It has a genetic component but is a response to environmental cues. Migrating birds face numerous hazards, and conservation of migrants relies upon efforts made in a number of countries. Birds use a wide range of navigational cues to facilitate their movements and in some cases have excellent spatial memories.

Questions for discussion

1. How are conservation efforts for migrant birds likely to differ from those of non-migrants?
2. How do birds find their way?
3. How is migration controlled?

CHAPTER 4

Eggs, nests, and chicks

The avian egg is a miracle of natural engineering

Noble S. Proctor and Patrick J. Lynch, 1993

Eggs, particularly perhaps those of domestic fowl, are something that we are probably all very familiar with, but just how much do we know about them? In this chapter, I want to consider the egg from its conception, through laying and incubation, to hatching.

Chapter overview

4.1 Sex and the gonads of birds
4.2 The egg
4.3 Egg shell coloration and patterning
4.4 Incubation and the developing embryo
4.5 Hatching
4.6 Chicks

4.1 Sex and the gonads of birds

Birds reproduce sexually; a male and a female copulate, sperm are transferred, an egg is fertilized (internally), and, assuming that all goes well, a new bird is the eventual result. Genetically this new individual will be a combination of its parents, having inherited half of its genetic material from each of them.

As humans, one of the chromosomes we inherit from our father determines our sex. Human sex chromosomes come in two types, labelled X and Y. Our fathers (like all male mammals) have paired sex chromosomes consisting of one X chromosome and one Y chromosome and when their germ cells divide at meiosis they produce gametes (sperm) that have either an X or a Y. Male mammals are therefore described as the heterogametic sex. But our mothers (and other female mammals) are homogametic, i.e. they have a pair of X sex chromosomes and so their gametes are all the same (X). This of course means that the sex chromosome that we inherit from our mother will always be an X chromosome, but from our father we may inherit either an X or a Y. If our paternal sex chromosome is a Y then we are male, if it is an X we are female. Birds too have one heterogametic sex and so sex is determined in the same way, but with two notable differences. The first is a difference in terminology—instead of X and Y chromosomes birds have W and Z chromosomes. The second difference is perhaps biologically more significant; whereas in mammals sex is determined by the material inherited from male parents, in birds it is determined by the material coming from the female because it is the female that is the heterogametic sex having a ZW pair of sex chromosomes (males have two Z chromosomes).

Chromosomes consist of a double helix of DNA, specific sections of which act as templates for the production of proteins. These protein-coding sections are what we refer to as genes. Each individual gene always codes for the same protein or part of a protein (some are the product of a number of genes working together) and so the expression of a particular gene will always affect the same character of the organism's behaviour, physiology, development, etc. The observable effects of the expression of genes are an organism's phenotype, but we should not think of phenotype (be it behaviours observed, physiological characteristics, or plumage colour) as being solely determined by genotype. Phenotypes also depend upon the environment in which they are expressed (see below).

The genes found on the Z and W chromosomes are referred to as being sex linked. For example, only a female bird can have any of the genes that are found solely on the W chromosome, so these genes are linked to being female. We are only beginning to understand the specific details of the impact of sex-linked genes upon the development and life of the individual, but I would like to mention briefly one exciting example of the clear link between a sex-linked gene and sex-specific development.

Male birds of a number of species sing a courtship song to attract a mate and as a form of resource defence (behaviours that we will discuss in more detail in Chapter 5). Singing is triggered by a wide range of environmental stimuli (by environment here I mean both the environment external to the bird, i.e. the physical and social environment, and the bird's internal environment, i.e. the particular hormones in its bloodstream, etc.), but song production is under the control of a specific cluster of brain cells termed the high vocal centre (HVC). The HVC of male songbirds is significantly more developed than that of female birds and this difference is initiated at an early stage in the development of the brain. But what exactly triggers this sexual difference in the neurological basis of song has been a mystery. Recently, however, Xuqi Chen and colleagues have reported that HVC development in the brain of the male Zebra Finch *Taeniopygia guttata* is temporally linked

> **Flight path:** development and control of bird song. Chapter 5.

to the expression of a Z-linked gene which codes for the protein tyrosine kinase receptor B (trkB) which acts as a receptor for a neurotransmitter termed BDNF (brain-derived neurotrophic factor). BDNF in turn is known to be involved in the differentiation of brain cells and in the development on the HVC in particular. As we might reasonably expect, male birds have far higher trkB levels, having twice as many trkB Z-linked genes than females. Thus a direct link between a sex-linked gene and a sex-specific behaviour has been established (although the exact mechanism for its effect has yet to be confirmed).

Although the details have yet to be determined, it can be assumed that the basic genetic difference between male and female birds also in some way determines the differentiation of their tissues into gonads that are either male (testes) or female (ovaries).

Males have a pair of internal testes which produce sperm. These are enlarged during the breeding season but shrink away to almost nothing during the rest of the year. Females of most bird species have only one developed ovary (usually the left one), although there are exceptions and the females of some species of bird of prey do have two ovaries (although they may not both be functional). As in the case of the male testes the female ovary is far larger during the breeding season than it is during the rest of the year. Presumably this seasonal development of gonad tissue is of benefit to birds that have to balance their body mass carefully to maximize their flight efficiency.

Birds on the whole do not possess an external sex organ such as the mammalian penis, and copulation is usually a very brief affair lasting just seconds, but can last for 25 minutes in the case of the Aquatic Warbler *Acrocephalus paludicola*. Both males and females have a cloaca, and during copulation the male places the opening of his cloaca over that of the female and ejaculates directly into her. In the passerines the male cloaca does swell and protrude during the breeding season and this may serve to increase the efficiency of sperm transfer; in a similar way some species of duck and members of a small number of other families develop a penis-like protuberance that is involved in copulation.

After transfer, sperm swim from the area of the female cloaca towards the top of the oviduct, to the infundibulum and ovary, where the mature ovum is fertilized (Figure 4.1). In some species females have special storage organs around the junction of the vagina and uterus where sperm may be retained in a viable state for some days or even weeks prior to their release, which is timed to coincide with ovulation. The significance of this is discussed in Chapter 5.

4.2 The egg

In their excellent *Manual of Ornithology*, Nobel Proctor and Patrick Lynch refer to the egg of a bird as '...*a miracle of natural engineering. Light and strong, it provides*

Key reference

Chen, X., Agate, R.J., Itoh, J., and Arnold, A.P. (2005) Sexually dimporphic expression of trkB, a Z-linked gene, in early posthatch zebra finch brain. *Proceedings of the National Academy of Sciences, USA* **102**, 7730–7735.

Flight path: seasonal modification of tissue mass, flight efficiency, and migration. Chapters 2 and 3.

Flight path: sperm storage and breeding strategies. Chapter 5.

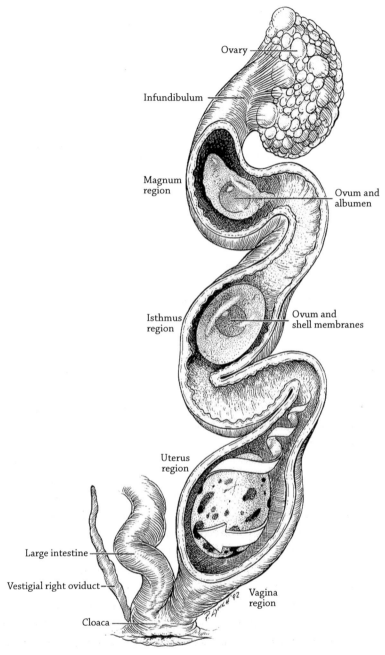

Figure 4.1 The oviduct and the formation of eggs. After its release from the ovary the egg typically takes 24 hours to develop fully: spending around 30 minutes in the upper area of the oviduct (the infundibulum); around 3 hours in the magnum region where it is coated with albumen; 1 hour in the isthmus region where shell membranes are deposited; and up to 24 hours in the uterus where the hard outer shell and associated pigmentation is laid down. From Proctor, N.S. and Lynch, P.J. (1993) *Manual of Ornithology: Avian Structure and Function*. Yale University Press, New Haven.

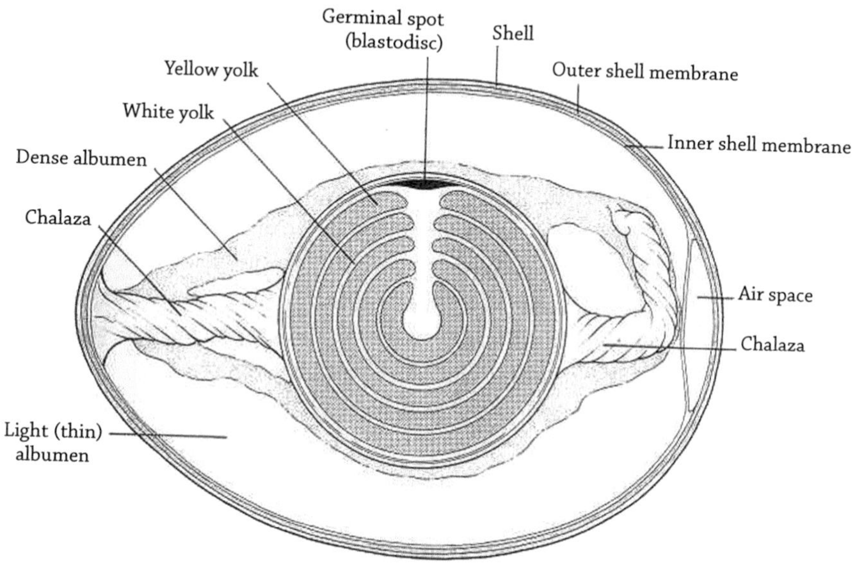

Figure 4.2 The internal anatomy of a generalized birds egg. From Proctor, N.S. and Lynch, P.J. (1993) *Manual of Ornithology: Avian Structure and Function.* Yale University Press, New Haven.

everything a developing bird embryo needs.' And this quote does I think sum up an egg perfectly.

Essentially the egg is a zygote, a fertilized ovum, sitting in a relatively huge store of nutrients and encased in a protective shell. But as Figure 4.2 illustrates, there is rather more to it than that.

At the centre of the egg is the yolk, rich in fats, proteins, and other nutrients. The yolk may comprise as much as 70% of the content of an egg in the case of the Brown Kiwi *Apteryx australis*, or as little as 20% in the case of small passerines. When we crack open an egg to cook it the yolk appears to be a relatively uniform yellow body, but in fact if prepared properly its real structure can be seen and it is revealed as being a series of alternating layers of nutrient-rich yellow and less rich white yolk. These layers are built up over a period of days as the yolk is formed, and the difference in their colour is an indication of differences in nutrient availability (layers laid down at night are less nutrient rich). So, just as we can age a tree by its growth rings we can determine yolk age (or at least measure the time taken for its development) by counting the layers within it. The developing embryo (beginning as a group of germ cells termed the blastula) sits on top of the yolk. Specifically it sits on top of a column of white yolk and because white yolk is less dense than yellow yolk this column ensures that as the egg rotates the embryo (and the white yolk under it) floats to retain its position above the yolk mass rather than underneath it. The yolk is contained within the vitelline membrane to maintain its integrity and is held in place by the chalazae,

Key reference

Proctor, N.S. and Lynch, P.J. (1993) *Manual of Ornithology: Avian Structure and Function.* Yale University Press, New Haven.

Flight path: yolk volume correlates with chick developmental strategy. Chapter 5.

gelatinous albumen structures which allow the aforementioned yolk rotation but otherwise limit its movement.

The yolk is surrounded by the albumen, commonly referred to as the white of the egg because of the transformation that it undergoes on cooking; changing as it does from an almost transparent gel to a white solid. If you fry an egg 'sunny side up' you will notice that there are two layers of albumen, a dense inner layer in contact with the yolk and a less dense outer layer. Consisting of around 90% water and 10% protein, the albumen is the water reserve for the developing chick. The albumen layer also acts as a physical cushion, protecting the embryo from sudden jolts as the egg moves, and as an insulating membrane reducing the cooling of the yolk and embryo when incubation is interrupted.

The outer shell of an avian egg is composed of a number of discrete layers, the most obvious being the hard outer layer of calcium carbonate crystals arranged in a lattice of flexible collagen fibres. This gives the shell the strength that it needs to bear the weight of an incubating parent and the resilience required as it is jostled against other eggs in the clutch and against its surroundings (remember that not all eggs are laid into a comfortable nest—some such as those of cliff-nesting seabirds are laid directly onto unyielding rock). Beneath this outer layer are two flexible inner membranes to which the brittle outer layer adheres, further contributing to the overall stability of the shell. The innermost shell membrane is in direct physical contact with the vascularized chorioallantoic membrane through which embryonic respiration takes place. To facilitate the transport of gasses into and out of the egg, the shell has permanently open pores allowing the inner membranes of the egg to communicate directly with the external environment.

Most birds lay eggs at the rate of one per day for a fixed period (depending upon clutch size, the determination of which is discussed in Box 4.1). There are of course exceptions, and some birds lay every other day, and as an extreme the Masked Booby *Sula dactylatra* lays the two eggs in its clutch 6 or 7 days apart. The reason for the interegg interval probably relates in part to the time it takes for an egg to be produced (see Figure 4.1) and the need on the part of female birds to maintain flying efficiency—carrying one almost fully formed egg is probably a sufficient burden in the context of aerodynamics.

Female birds do, it seems, have some control over the size and quality of the eggs that they lay. Elisabeth Bolund and her colleagues have demonstrated that female Zebra Finches increase the volume of the eggs that they lay and increase the carotenoid content of their yolks when they are paired with a low quality mate. In essence they lay better eggs. This is presumably an attempt by them to off-set the poor genetic quality (low attractiveness) of their mate by giving their offspring a bit of a head start.

Key reference

Bolund, E., Schielzeth, H., and Forstmeier, W. (2009) Compensatory investment in zebra finches: females lay larger eggs when paired to sexually unattractive males. *Proceedings of the Royal Society B: Biological Sciences* **276**, 707–715.

Box 4.1   Clutch size

Female albatrosses invariably lay a single egg whilst the clutch of a Blue Tit *Cyanistes caeruleus* can have as many as 17 eggs in it. There is clearly therefore considerable variability in clutch size. General patterns of variation in clutch size have been described, and interpretation of these does allow the formulation of some quite straightforward explanations for the phenomenon. For example, the clutches, and therefore broods, of altricial species whose young are highly dependent upon their parents tend to be smaller than those of precocial species (with young that are quite independent). It seems clear that in these cases the ability of the parents to feed, brood, and protect the chicks is a key determinant of clutch size. It has also been noted that species utilizing open nests have smaller broods than those with a more secure cavity nest. Presumably the increased protection from predation afforded by a cavity nest is important, but it is also possible that open-nest broods are smaller to enable faster fledging and so minimize predation by cutting short the risky period during which chicks are in the nest.

It is also clear that there is a heritable component to clutch size; after all, artificial selection for increasing 'clutch size' has allowed humans to develop varieties of domestic fowl that are able to lay an egg a day almost all year round.

Females of some species are extremely limited in terms of the variation in clutch size that they can achieve (e.g. female Spotted Sandpiper *Actitis macularis* always lay four eggs, the implications of which are discussed in Chapter 5). On the other hand the females of many species are able to vary the number of eggs that they lay between nesting attempts/ seasons. In these cases the question 'How big should a clutch be?' is one that has been a focus of considerable interest to ornithologists.

Early observations of clutch size took the view that the number of eggs laid by a female should be the same as the maximum number of young that could be successfully reared, and that this was likely to depend upon the food resources available. Field-based observations suggest that in fact clutches are most often slightly smaller than would be expected. This disparity between theoretical and actual clutch size can be explained if, rather than considering a single breeding season, the prediction is made that the optimal clutch size will be the one that allows a female to maximize her productivity over her lifetime. In effect, by producing slightly fewer eggs in any one clutch, females are ensuring that they will have the reserves/survivorship potential to lay further clutches in future.

Evidence for this kind of subtle brood manipulation is provided by Goran Högstedt who noted that the clutches laid by individual females of the Swedish Black-billed Magpie *Pica pica* population that he studied varied between five and eight eggs. To test the idea that these females somehow 'knew' what the optimal number of eggs to lay was, he manipulated brood sizes over three breeding seasons. Some broods were not manipulated and these females were allowed to rear all of the chicks that they hatched.

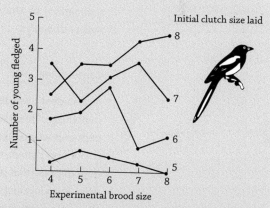

Figure 4.3 Experimental manipulation of brood size demonstrates that female Magpies *Pica pica* consistently lay the number of eggs that their territory can support. From Högstedt, G. (1980) Evolution of clutch size in birds: adaptive variation in relation to territory quality. *Science* **206**, 1148–1150.

Some females had their brood size reduced, and the newly hatched chicks that were removed from them were used to increase the broods of others. The data presented in Figure 4.3 show that in all cases the optimum clutch size (the one resulting in the most chicks still alive at the point of fledging) was the same as that laid by the female. Through further work Högstedt was able to show that the main factor determining clutch size in this population was territory quality (food availability and predation risk), and that at some level females were able to relate territory quality to clutch size and consistently lay the optimum clutch. This story will be returned to in Chapter 5 when we consider magpie parenting behaviour.

Figure 4.3 also highlights some of the consequences of not laying the optimal clutch; lay too many or too few eggs and productivity will decline as a result of the twin pressures of starvation and predation.

4.3   Egg shell coloration and patterning

Eggs are quite simply beautiful objects. Their form is particularly pleasing to the human eye for some reason, and the diversity of their colours and patterns is quite amazing. There are two components to egg shell colour—the base colour (usually white, pale brown, or pale blue, but deep reds, blacks, and greens are also found), and the pattern (streaks, spots, or blotches in a range of colours that are usually, but not always, darker than that of the base). The natural beauty of eggs sparked a frenzy of egg collecting and their study (oology) during the Victorian period. Over the period of their academic study a host of explanations as to the significance of their colour and pattern have been proposed. These are very well summarized by Underwood and Sealy in their contribution to the monograph *Avian Incubation*.

Key reference

Underwood, T.J. and Sealy, S.G. (2002) Adaptive significance of egg colouration. In: Deeming, D.C. (ed). *Avian Incubation*. Oxford University Press, Oxford.

Camouflage

As a student I was told that the colour of an egg was an adaptation to facilitate camouflage. This, it was explained, was why white eggs were laid by cavity-nesting birds and why patterned eggs were laid by birds that had open nests (with the colour and pattern matching the habitat around the nest). There may be some truth to this in that some eggs are camouflaged—personally I have often had great difficulty in locating the nests and greenish/brown mottled eggs of Oystercatcher *Haematopus ostralageus* despite their being laid on open ground in nothing more elaborate than a scrape. In my local Oystercatcher population the eggs are 'lost' amongst the gravel mix of the upper shore on which they are laid (Plate 12).

However, not all eggs that are laid in the open are camouflaged and not all cavity nesters lay white eggs. So whilst this explanation might explain the coloration of a proportion of eggs, it is insufficient as an explanation of the coloration of all eggs. The camouflage explanation for egg colour is one of a family of explanations which ascribe a signal function to the appearance of the egg (the cryptic pattern is effectively a dishonest signal to a predator). Others in this family of explanations include the evolution of patterns to facilitate egg recognition by incubating parents and/or the evolution of patterns to signal female quality (presumably to male birds).

Egg mimicry

In an attempt to stay one step ahead of the competition, some brood-parasites such as the Cuckoo *Cuculus canorus* lay eggs that accurately mimic those of their host, thereby making their discrimination and rejection harder. At the population level, female cuckoos lay eggs of a range of colours and patterns but each individual female lays only one type and preferentially parasitizes only one or a small number of the wide range of hosts available. Those that she chooses are the ones with eggs most similar to her own. This host specificity is thought to be genetically controlled and sex linked to the W chromosome. There is experimental evidence to support the utility of this egg mimicry in that in a range of host–parasite pairs, hosts have been found to reject a higher proportion of non-mimetic than mimetic eggs.

Egg recognition

Recognition of one's eggs is likely to be important in two key situations: when trying to find them in a crowded colony; and when trying to separate them from others laid alongside them within a clutch. In the former case it is important that incubating birds are able to discriminate between their own eggs and those of their neighbours in a colony. In many colonies the eggs of open-nesting birds are laid in very close proximity to one another and although birds returning to incubate are likely to gain information from a variety of cues relating to nest position in the colony and perhaps directly from their partner, if it is in attendance at the nest, the ability to recognize ones own eggs is clearly important. It has been shown experimentally that guillemots are able to discriminate between their own egg and those of their neighbours on a crowded cliff ledge because they recognize the colour and blotch pattern of their own particular egg.

Egg recognition is also important in the reduction of the impact of brood-parasitism—the behaviour by which a female lays her eggs in the nest of another bird. Brood-parasitism can occur between individuals of the same species (when it is more usually referred to as egg dumping), but is probably better known as an interspecies activity. Brood-parasitism in the traditional (interspecific) sense is a comparatively rare breeding strategy found in less than 1% of bird species but having evolved in a broad range of taxa, notably the cowbirds (Icteridae), whydahs (Viduidae), honeyguides (Indicatoridae), and the cuckoos (Cuculidae). It is also exhibited in a number of species which have precocial young (quite independent at hatching), often as part of a wider reproductive strategy that does involve incubation of their own young. So, for example, brood-parasitism amongst the various species of duck (Anatidae) is widespread and, with one exception, all of the species involved do most commonly rear their own young. The exception in this case is the Black-headed Duck *Heteronetta atricapilla* which is an obligate brood-parasite (i.e. it never rears its own young) and has been recorded laying its eggs in the nests of 18 different species, including gulls, ibises, herons, coots, rails, and even birds of prey.

Egg dumping is common, for example, amongst some species of colonially nesting African weaver birds (Ploceidae), the females of which occasionally lay an egg in the nest of one of their neighbours. This might be to their advantage as both an insurance of their output against the risk that their own nest will suffer predation and as a means by which they can increase their total productivity. Of course from the viewpoint of the recipient of the dumped egg, investment in the incubation and rearing of the young of another pair is not an advantage as it reduces the investment that can be made in their own young. Weaver egg patterns are very diverse, perhaps as an adaptation to increase the ability of females to recognize their own eggs and enable them to remove those of their competitors. Some weaver bird species are the hosts of the Diederik Cuckoo *Chrysococcyx caprius*, a brood-parasite in the traditional interspecific sense of the term, laying its eggs in the nest of another species and leaving its young to be raised by these unwitting foster parents. It is possible, therefore, that egg recognition by weavers is also a strategy to minimize the effect of this behaviour. Cruz and Wiley provided compelling evidence to support this argument in their observations that a population of Village Weaver *Ploceus cucullatus* introduced to the Caribbean island of Hispaniola lost most of their egg recognition ability during the 200 years of brood-parasite-free existence that they enjoyed prior to the establishment of a population of Shiny Cowbird *Molothrus bonariensis*. After just 16 years of coexistence with this brood-parasite, the discriminatory ability of the weavers had almost completely returned.

> **Flight path:** brood-parasitism is just one of a wide range of reproductive strategies. Chapter 5.

> **Key reference**
>
> Cruz, A. and Wiley, J.W. (1989) The decline of an adaptation in the absence of presumed selection pressure. *Evolution* **43**, 55–62.

Box 4.2   When does it pay hosts to accept brood parasites?

Natural selection seems to favour those host birds able to recognize and eject the eggs of brood-parasites. And numerous examples of the 'arms race' between the egg mimicry ability of the parasite and egg recognition abilities of the host have been described. But it seems that there may be instances when the parasite can get the upper hand and it will pay the host to accept the cost of rearing the 'cuckoo in the nest'.

Jeffery Hoover and Scott Robinson have carried out an elegant series of experiments involving the relationship between Brown-headed Cowbirds *Molothrus ater* and their Prothonotary Warbler *Protonotaria citrea* hosts. In this system, unlike that of other brood-parasites, hosts raise the alien chick alongside their own brood—paying an energetic cost and producing weaker chicks as a result.

The researchers wanted to investigate two interesting hypotheses: (1) that the cowbirds may 'farm' their hosts, i.e. induce parasitism opportunities; and (2) that the warbler hosts may be 'intimidated' into accepting parasitism by the consequences of not doing so. In essence they suspected the parasites of mafia-like behaviour—punishing those warblers that ejected their eggs by returning to destroy their clutches.

To test their ideas, Hoover and Robinson established a nest box breeding population of warblers in an area frequented by cowbirds and carefully monitored nesting attempt outcomes. They noted parasitism rates and manipulated parasitized nests. Some they allowed to develop naturally but from some they removed the cowbird egg. Some, but not all, of these manipulated nests then had their entrance hole reduced to permit warbler access whilst excluding cowbirds.

Look carefully at the results of the experiment (Figure 4.4). Treatments 3, 4, and 5 show that in ideal conditions (no parasitism or egg removed but no possibility of a return to the nest by the cowbird) a Prothonotary Warbler pair might expect to raise three or four chicks per breeding attempt, and that these nests were rarely predated. Treatment 3 also reveals a cost to the warblers of accepting the parasite in that warbler productivity is lower in these nests than in the nests of treatments 4 and 5. The cost paid by accepting the parasite (about one chick) is, however, far lower than that paid by birds that reject the egg (treatment 1)—who on average raise only one of their chicks. These nests suffer high predation rates (as do unparasitized eggs) and the most likely predator in this case is the cowbird.

Hoover and Scott interpret their findings as follows: When they reject a cowbird egg, warblers induce a

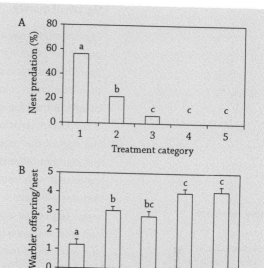

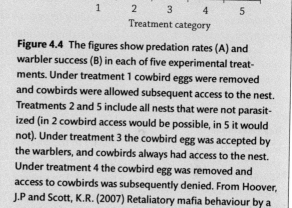

Figure 4.4 The figures show predation rates (A) and warbler success (B) in each of five experimental treatments. Under treatment 1 cowbird eggs were removed and cowbirds were allowed subsequent access to the nest. Treatments 2 and 5 include all nests that were not parasitized (in 2 cowbird access would be possible, in 5 it would not). Under treatment 3 the cowbird egg was accepted by the warblers, and cowbirds always had access to the nest. Under treatment 4 the cowbird egg was removed and access to cowbirds was subsequently denied. From Hoover, J.P and Scott, K.R. (2007) Retaliatory mafia behaviour by a parasitic cowbird favours host acceptance of parasitic eggs. *Proceedings of the National Academy of Sciences USA* **194**, 4479–4483.

high price—a visit from the cowbird and predation of their clutch. It therefore pays them to accept the egg in a scenario that is somewhat like a mafia protection racket! By predating warbler nests, the cowbirds act as 'farmers' inducing the laying of second clutches which provide new parasitism opportunities.

Signals of quality

Egg coloration may also be a signal used by males to judge the quality of a female and of her offspring. Juan Moreno and coworkers have shown that the eggs of a Spanish population of Pied Flycatcher *Ficedula hypoleuca* vary in the intensity of their blue-green coloration (measured in terms of the intensity of their reflectance of light in the blue-green region of the visible spectrum); eggs are similar within a clutch, but variation between clutches is marked. The blue-green coloration of the eggs of this species is a result of the deposition in the matrix of the shell of the pigment biliverdin. This pigment has antioxidant properties and is positively correlated with immunocompetence in adult birds. It is likely that only healthy females will have a sufficient surplus of biliverdin to produce really blue-green eggs. Egg colour could therefore be a signal of female health status or quality; but what evidence is there that males actually take note? To explore this question the researchers made observations of male provisioning rates at nests of known egg colour type (they also cross-fostered clutches to ensure males really were responding to egg colour and not to laying female behaviour or other cues). They found that males made more provisioning visits to the nests containing the most colourful eggs so they did invest more in the chicks of healthy females (presumably these females produce healthy chicks). Thus it is possible that egg colour was used by males as a cue.

On the other hand, José Martínez-de la Puente and coworkers have demonstrated a correlation between egg shell patterning and a measure of low female quality in the Blue Tit *Cyanistes caeruleus*. Blue Tit eggs are white with red-brown spots that are the result of the incorporation of protoporphyrins in the matrix of the shell. The presence of these pigments in the shell seems to be a consequence of elevated protoporphyrin levels in the laying female. These pigments are oxidants rather than antioxidants, and at elevated levels they can indicate and even cause poor health in the adult bird.

Specifically the researchers have shown that females in poor body condition, indicated by their having a higher cellular level of a protein HSP70 (heat shock protein 70) which is known to be linked to stress, laid spottier eggs. So perhaps increased patterning is a means by which females can rid their bodies of excess protoporphyrins and improve their own health status. There is no evidence currently that male Blue Tits respond either positively or negatively to this potential signal.

Pigments and shell quality

An alternative explanation of the significance of protoporphyrin pigments in passerine egg shells has been proposed by Andrew Gosler and coworkers working with the eggs of a population of Great Tits *Parus major*, which like Blue Tits have a white egg patterned with brown spots. Birds in their study population (at Wytham

Key reference

Moreno, J., Morles, J., Lobato, E., Santiago, M., Tomás, G., and Martínez-de la Puente, J. (2006) More colourful eggs induce a higher relative paternal investment in the pied flycatcher *Ficedula hypoleuce*: a cross fostering experiment. *Journal of Avian Biology* **37**, 555–560.

Key reference

Martínez-de la Puente, J., Merino. S., Moreno, J., Tómas, G., Morales, J., Lobati, E., García-Fraile, S., and Martínez, J. (2007) Are eggshell spottiness and colour indicators of health and condition in blue tits *Cyanistes caeruleus*? *Journal of Avian Biology* **38**, 377–384.

Woods near Oxford, UK) laid rounder spottier eggs if their territories were low in calcium, and more elliptical less spotty eggs if their territories were high in calcium (Plate 13).

The significance of these observations seems to be that the brown spots are coincident with thinner (less calcium-rich) areas of the shell, and spottier eggs therefore had generally thinner shells. A thinner shell is a weaker shell and this might explain the recorded difference in egg shape because a rounder egg is stronger than an elliptical one. This difference in shell quality can be easily explained—female birds must obtain calcium from their environment, and in a calcium-poor environment they will produce poorer eggs. But what about the spots? Being coincident with thinner areas of shell it is presumed that they play an important structural role and act as strengthening agents.

Of course it is possible that the potential health status signal and egg shell strengthening roles of protoporphyrins are not mutually exclusive—birds low in calcium might be in generally poor condition, producing poor quality eggs.

Key reference

Gosler, A., Higham, J.P., and Reynolds, S.J. (2005) Why are birds' eggs speckled? *Ecology Letters* **8**, 1105–1113.

Box 4.3   Nests

Eggs are laid in nests. However, the most simple nest of all is really no more than a 'nest-site'—the largely unmodified bare ground/wood onto which an egg is placed. Many species of wader, for example, lay their eggs directly into a shallow depression or scrape on the ground (see Plate 12). Similarly cliff-nesting auks such as Guillemot *Uria aalge* and Razorbill *Alca torda* lay their eggs directly onto what seem to be the narrowest of rock ledges. Nests are placed within cavities or burrows; stuck to vertical surfaces with mud, faeces, and saliva (Plate 14); they float 'tethered' to emergent vegetation; they hang over water; they are placed in vegetation at ground level, in shrubs, and in tree-tops (Plate 15). Some are so small that females straddle them rather than sitting on them when incubating eggs, and others are so large that their weight can bring down the tree in which they have been built (the huge colony nests of some social species of African weaver bird for example). In fact the variety of nests, nest locations, and nesting materials is so enormous that it is beyond the scope of an introductory text such as this and I would recommend that the interested reader consult Mike Hansell's excellent book *Birds Nests and Construction Behaviour*.

The primary function of a nest is to provide a safe place in which eggs can be laid and incubated, and where, in the case of altricial species, chicks can be safely reared. Domed and enclosed nests afford protection from the elements and may help to maintain a suitable environment for developing eggs and chicks. Many birds attempt to maintain nest hygiene by defecating over the lip of the nest, and the chicks of many of the passerines produce discrete faecal pellets which are collected and ingested or carried away by their parents. (This copraphagy may seem unpleasant but has been shown to make a significant contribution to the nutrition of some breeding birds.) To further maintain nest health many species regularly add fragrant green vegetation to their nests. It has been suggested that the volatile chemicals given off by these plants may act as insecticides or arthropod deterrents and by including them birds may be able to limit the impact of harmful nest parasites. Although data to support this idea are not yet available it has been shown by Adèle Mennerat and her coworkers that Blue Tit chicks raised in nests with aromatic plant fragments have higher haematocrit levels than those raised without them. Haematocrit levels are related to the oxygen-carrying capacity of the blood and are an indicator of general health, so it does seem that there is a benefit to inclusion in the nest of this material in that it may act as a 'drug' enabling birds to better withstand parasite attack.

References

Hansell, M. (2000) *Birds Nests and Construction Behaviour.* Cambridge University Press, Cambridge.

Mennerat, A., Peret, P., Bourgault, P., Blondel, J., Gimenez, O., Thomas, D.W., Heeb, P., and Lambrects, M. (2009) Aromatic plants in nests of blue tits: positive effects on nestlings. *Animal Behaviour* **77**, 569–574.

4.4 Incubation and the developing embryo

In order for the embryo inside to develop, eggs must be kept warm. In fact the eggs of most species require a fairly constant temperature of around 38°C (100°F). Relatively few birds experience such temperatures in their environment and for those that do environmental temperatures are likely to fluctuate to too great an extent to make ambient incubation a possibility. Australasian megapodes do of course practise environmental incubation, burying their eggs under mounds of earth and organic matter, which as it decays releases the steady heat that they require. The majority of species, however, practise direct incubation, with parent birds 'sitting' on the eggs until they hatch (Plate 16). In many cases both sexes share incubation duties, but there are numerous examples of female-only incubation and more rarely examples of male-only incubation.

Incubation periods vary greatly; the eggs of some species of African weaver bird (Ploceidae) hatch in just 9 days, whereas those of some penguin species take 65 days, and those of kiwis can take an amazing 85 days to hatch.

A few days before the onset of incubation, physiological changes occur in the parent bird; levels of the hormone prolactin circulating in the bloodstream increase and at the same time levels of circulating testosterone decrease. This hormone shift is thought to be a trigger for the onset of incubation and parenting behaviours, and for a reduction in the performance of behaviours related to territoriality and courtship. All of these behaviours will be further discussed in Chapter 5, but at this stage I do want briefly to describe the development in the incubating bird of a specialized 'organ' to facilitate the transfer of body heat to the egg. Feathers on the lower breast and belly are lost (they drop out or are pulled out, and may then be used in nest insulation) and the bare skin beneath them swells with fluid. The blood vessels beneath this skin dilate, thereby increasing blood flow to this region. This brood patch as it is termed (Plate 17) is concealed by the contour feathers of a non-incubating bird, but as it settles onto the nest the feathers are drawn back such that the hot skin of the patch makes direct contact with the eggs. By raising

Flight path: Reproductive strategies vary between species. Chapter 5.

and lowering their body over the eggs birds can control their temperature and in extremes of heat some species stand above the eggs to give a parasol effect—if this is insufficient they make a trip to a local water supply, wet their breast feathers, and then return to drip cooling water directly onto the eggs. The brood patch is a temporary feature of incubating birds and, once the eggs have hatched and the developing brood no longer need to be warmed, it shrinks away and the feathers re-grow when the bird next moults.

You will recall that at the time of laying, the embryo within the egg is little more than a cluster of cells on the surface of the yolk (Figure 4.1). As incubation progresses these cells repeatedly divide and differentiate as the embryo becomes recognizably a bird. To facilitate the uptake of nutrients from the yolk and to permit the exchange of gases with the environment, the developing embryo produces two extraembryonic membranes. One of these, the yolk sac, is a vascularized sheath surrounding the yolk and acting as a kind of external stomach, allowing the embryo to absorb nutrients directly from the yolk. Shortly before hatching the yolk sac is absorbed into the body of the chick and it may serve as a nutrient reserve to enhance its chances of survival during the first few days after leaving the egg. As embryonic development progresses, the chick absorbs the calcium needed for bone growth from the shell of the egg, thereby weakening it and presumably increasing the ease with which the chick will be able to crack it when the time comes.

The second membrane, the chorioallantois, develops between the inner surface of the egg shell, eventually covering most of it. This highly vascularized membrane provides for the transport of oxygen into and carbon dioxide and excess water vapour out of the embryo. As we have already seen, these gases enter and leave the chorioallantois and the egg via pores in the outer shell.

Box 4.4   Birds, eggs, and agricultural chemicals

The second half of the twentieth century began with a global expansion in and intensification of agriculture. This involved a global expansion in the application of new pesticides and herbicides, designed to increase crop yields by decreasing populations of crop pests and competitors. There are even pesticides designed to target bird species directly in soft fruit-growing areas. However, from the ornithologist's perspective, perhaps the most notorious of these chemicals were not those intended to control birds but those which had an indirect impact upon them.

During the 1950s and 1960s marked declines in numbers of birds of prey were recorded throughout the countries of the developed world. Adult birds were found to have died for no apparent reason (they were not the victim of trauma for example), and clutches of eggs were found broken and/or abandoned. In 1970, Derek Ratcliffe published a keystone paper in the *Journal of Applied Ecology* in which he demonstrated that the main cause of the population declines was breakage of eggs in the nest, a phenomenon that he showed to have arisen quite suddenly in the mid/late 1940s. After making measurements of shells laid during the period of the decline, and comparing them with collected eggs from earlier in the century, he proved that the increase in egg breakage coincided with a sudden change in the quality of egg shells, in simple terms the shells of the eggs had become thinner (Figure 4.5). Ratcliffe's paper did more than just demonstrate the incidence of thinning; it also provided correlational evidence to support the idea that the most likely cause of the thinning was an accumulation in the birds concerned of residues of organochlorine pesticides such as DDT which had come into widespread use in the period 1946–1950. Subsequent experimental work has confirmed that the breakdown products of organochlorines (DDE from DDT, HEOD from dieldrin and aldrin, and PCBs) do all impact upon the calcium metabolism of birds, thereby directly causing egg shell thinning.

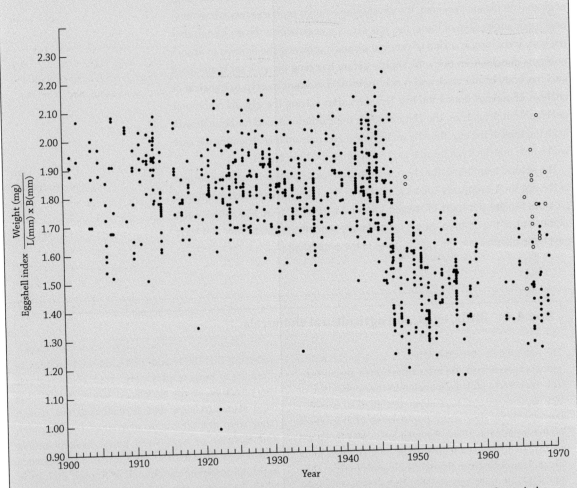

Figure 4.5 Note that eggshell thickness for UK Peregrine *Falco peregrinus* remained fairly constant during the period 1900-1945 but that during the postwar years (1945-1970) shells became considerably thinner on average. This was coincident with increased use of DDT. Adapted from Ratcliffe, D. (1970) Changes attributable to pesticide in egg breakage frequency and eggshell thickness in some British birds. *Journal of Applied Ecology* 7, 67–115.

Having identified organochlorine pesticides as a main factor in the decline of bird populations, many governments legislated against the general use of organochlorines and against the use of DDT, aldrin, and dieldrin in particular (they were banned in the UK in 1986 for example). But has it done any good? Well thankfully it has. Take, for example, the case of the Merlin *Falco columbarius*, a small raptor common throughout North America and northern Europe and a bird that is known to have suffered egg shell thinning and reduced breeding success throughout its range. Ian Newton and colleagues have shown that in the case of the British merlin population, legislation has probably saved them from the brink of extinction. By the 1980s their numbers had fallen to just 500 or so pairs in the whole of the British Isles; just 10 years later the population had more than doubled to an estimated 1,300–1,500 pairs. This increase coincided with both a fall in the detectable levels of organochlorines such as DDE in eggs from failed broods and an increase in an index of egg shell quality approaching the pre-1946 level (Figure 4.6).

So have we as a society learned a valuable lesson from this potentially disastrous episode? The answer to that question is a qualified yes. It took decades to recognize the issue relating to organochlorine use, decades for positive remedial action to be taken, and only now decades later are we seeing the full benefits of those actions (and it should be remembered that high levels of these chemicals in the environment are still having their effect in a number of parts of the world). But as a society we continue to rely upon the introduction of chemicals into the environment to solve our agricultural problems.

Take, for example, the catastrophic demise of the worlds vultures during the opening years of the twenty-first century. Populations of some vulture species declined massively between 1990 and 2000; numbers of Oriental White-backed Vulture (*Gyps bengalensis*) on the Indian sub-continent fell by a staggering 95%, for example. Initial speculations suggested an epidemic specific to vultures might be

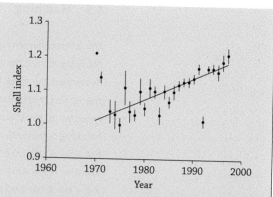

Figure 4.6 Shell index (thickness) values for the UK Merlin *Falco columbarius* have increased during the post-DDT era, almost returning to pre-DDT levels (index approximately 1.25). From Newton, I., Dale, L., and Little, B. (1999) Trends in organochlorine and mercurial compounds in the eggs of British Merlins Falco columbarius. *Bird Study* **46**, 356–362.

to blame, but very quickly diagnostic tests revealed that a pollutant was the causal agent. Specifically, in 2004 Lindsay Oaks and coworkers demonstrated that the anti-inflammatory drug Diclofenac was causing renal failure in the birds. This drug was in widespread use in the treatment of domestic livestock. When treatment was unsuccessful, livestock carcasses were simply left to be scavenged by the vultures and the birds ingested the drug and died as a result. By 2006, just 2 years later, India had announced its intention to ban such use of Diclofenac in an attempt to halt the decline of vulture populations and hopefully protect these birds from extinction. So it would appear that although we have not yet learned to avoid such mistakes, we have learned to identify them and respond positively to them more quickly.

Reference

Oaks, J.L. *et al* (2004) Diclofenac residues as the cause of vulture population decline in Pakistan. *Nature* **427**, 630–633.

4.5   Hatching

You will recall from Figure 4.2 that within the egg there is an air space. This is particularly significant during the period immediately prior to hatching. Once it has fully developed the chick begins to push with its beak against the membrane of the air space. This marks the beginning of hatching—a process that is thought to be triggered when the chick has quite simply outgrown its space. Specifically the onset of hatching coincides with increased hypoxia and hypercapnia, i.e. the chick is simply not able to take up enough oxygen, or expel enough carbon dioxide via the chorioallantoic respiratory system. In effect it needs to hatch before it suffocates. To assist it the chick has two temporary anatomical features related to hatching. One is an overly developed hatching muscle (more properly termed the complexus muscle) at the back of the head and neck which provides the extra strength needed to break out of the shell. The other is the egg tooth, a sharp process at the tip of the upper mandible of the beak that is used to pierce the air space membrane. Once the air space is broken into, the chick begins to use its lungs to breath air directly. The chick now uses the egg tooth to scrape the inner surface of the shell and, by a combination of scraping and pushing, it eventually makes a tiny hole. By repeating this process, and at the same time rotating the egg, the chick eventually weakens the shell sufficiently to cut of the cap and break out.

Interestingly the protoporphyrin pigments of brown speckled eggs that were discussed in an earlier section of this chapter may be important in hatching too. If you remember, there is a suggestion that these pigments play a strengthening role in thinner areas of shell. In fact the qualities of the pigment–shell matrix that strengthen it against forces from outside of the egg may in fact operate in reverse against forces from within—effectively making it easier for the chick to break the shell. In the Great Tit it has been noted that maximum areas of pigmentation coincide with the shoulder of the egg—the area first breached by the chick.

The process of hatching (or pipping as it is sometimes called) can take anything from a few hours in the case of a small passerine, to a few days in the case of some of the larger birds. Often the chick completes the process unaided, but there are examples of a helpful parent assisting in the final stages of the break-out. Having hatched, parental assistance is almost always essential. Newly hatched chicks are exhausted, wet, and extremely vulnerable to predators. As a minimum, parents brood chicks until they dry, but the extent of the care that they provide beyond that will vary from species to species.

4.6   Chicks

Newly hatched chicks demonstrate a range of levels of development. At one extreme, the chicks of Australasian megapodes (Megapodiidae) hatching from eggs that have been incubated for a prolonged period in a mound of compost or fermenting vegetation require no parental care. They hatch feathered and are able to fly almost at once. They are also able to thermoregulate and to forage for themselves. Such chicks are classed as being superprecocial. At the other extreme are the newly hatched chicks of the passerines. Hatching blind, naked, and helpless these altricial chicks rely entirely upon their parents for warmth, food, and protection. Interestingly, it has been suggested that such helplessness is only possible because passerines have evolved the ability to construct a complex nest which protects chicks from predators and from inclement weather.

Between these two extremes there are various grades of precocial/altricial development. In his review of the subject, Starck suggests that eight different classes of chick can be recognized: the superprecocial, three grades of precocial chick, semiprecocial and semialtricial, and two grades of altricial development which differ principally in the rate of their growth.

According to Stark's classification precocial chicks are mobile and sighted upon hatching and require varying degrees of parental care (Plate 18). For example, the chicks of ducks and pheasants follow their parents and are protected by them. They are initially down covered and so must be brooded by a parent to survive cold/wet weather, but like megapode chicks they are able to feed themselves from hatching. The precocial chicks of coots and rails are very similar to those of pheasants, but they are initially unable to find their own food and so must be fed by their parents (who place food in front of hungry chicks and demonstrate pecking behaviour to them).

The semiprecocial chicks of most gulls and terns are down covered and fed/brooded by their parents at the nest. However, when threatened they will leave the nest, which is typically quite exposed, and run/swim for cover, returning to the nest when it is safe to do so. Interestingly, but perhaps not surprisingly, the young of cliff-nesting Kittiwakes *Rissa tridactyla* differ from those of their gull cousins in that when danger threatens them they do not flee the nest (a bad thing to do if you live on a cliff face!). Instead they crouch in the nest cup and as a result of their cryptic plumage can be quite difficult to see. These chicks would be classed as being semialtricial.

> **Key reference**
>
> Stark, J.M. (1993) Evolution of avian ontogenies. *Current Ornithology* **10**, 275–366.

Summary

Eggs permit the external development of young, allowing female birds to maximize output without compromising flight. Females routinely lay optimally sized clutches, although clutch size, location of the nest, and incubation vary greatly from species to species. In some cases birds do not care for their own young, and such egg dumpers/cuckolds are involved in an evolutionary arms race with their hosts. As eggs and chicks, young birds are particularly vulnerable to predation and to the consequences of pollution.

Questions for discussion

1. Why do some species have precocial young whilst others have altricial young?
2. What pressures might come to bear in the determination of an optimum clutch size?

Reproduction

The dunnock's sex life is an arrangement of huge complexity.

Mark Cocker, 2005

In Chapter 4, I outlined the basic sequence of events from the laying of an egg to the hatching of a chick. In Chapter 5, I want to explore in more detail the behavioural processes leading to egg laying; territoriality and courtship; and the diverse mating systems exhibited by birds.

Chapter overview

5.1 Males and females are different
5.2 Mating systems
5.3 Courtship and mate choice
5.4 Bird Song
5.5 Raising a family

5.1 Males and females are different

Males and females are different. For example, in Chapter 2 we saw that as a result of the chromosomal difference between males (ZZ) and females (ZW) a chain of hormonal and developmental events eventually result in singing behaviour in mature male songbirds but not in mature females. Singing is a behaviour that we will return to later in this chapter. I also made the point in Chapter 4 that the gonads of males and females differ, and of course associated with that difference their gametes, the products of the gonads, also differ. Male gonads produce millions of sperm (the male gamete) at each ejaculation, only one of which is needed to fertilize the female gamete, the ovum or egg. It is significant that

Flight path: the relationship between the genetic make-up of the sexes and their behaviour. Chapter 4.

**Concept
Anisogamy**

Males and females have different sized gametes. Those of males (sperm) are small, mobile, and relatively inexpensive. Those of females (eggs) are relatively large, immobile, and expensive.

following ejaculation stores of sperm can be quickly replenished and so males are in theory capable of multiple matings and could sire large numbers of offspring during a period of reproductive activity. Sperm are therefore often thought of as being relatively inexpensive to produce and each individual sperm is in itself probably not a particularly significant investment on the part of the male. Eggs, on the other hand, are relatively large and they are relatively expensive to produce. Eggs are also in finite supply and so each of them has significant value to the female, representing as it does one of a very limited number of reproductive opportunities available to her. This fundamental difference in gamete size is termed anisogamy, and it is important because the reproductive strategies of birds (and other animals) are largely a consequence of it.

In basic terms we presume that all individuals seek to maximize their own reproductive output, by which we mean that they seek to pass on as many copies of their own genes as possible. We can therefore assume that individuals of both sexes behave in a way that will maximize their reproductive success in terms of the quantity of offspring produced and/or the quality of those offspring. As a result of anisogamy it is the case that males and females can probably maximize their reproductive output in different ways. Because a male can mate repeatedly, taking advantage of his easily replenished store of cheap sperm, we might reasonably assume that males can most easily increase their output by fathering as many young as possible. Females, on the other hand, cannot usually adopt the same strategy. They are constrained by a limited supply of eggs, and by the fact that there is a delay between successive ovulations. For a female, therefore, the most effective way to maximize reproductive output is for her to maximize the quality of her young. Females should therefore be expected to be choosy about their mates, seeking to maximize the quality of the contribution made by the male parent in terms of the quality of either his genetic material or the resources that he is able to provide.

One further consequence of the differences between male and female birds is that although actual sex ratios may be close to 1:1 (i.e. there are the same numbers of males and females of a species present in a population), operational sex ratios may depart from this significantly. This is because once a female bird has successfully mated she is likely to be unavailable for mating with another male because of the time she spends laying her fertilized egg and then, in the majority of cases, because she will invest time and energy in the successful incubation and rearing of her eggs and chicks. During this period of time the male is not similarly constrained and may be free to seek another mate. As a breeding season progresses, therefore, the operational sex ratio of the population will skew towards the relatively numerous males and away from the available females. In effect, therefore, females can be thought of as being the rarer sex. Charles Darwin recognized that this situation is the basis of the process of sexual selection, an important evolutionary force, by which the members of the rarer sex can be choosy about their mates, and those of the more common sex will be forced to compete at some level for access to mates.

Box 5.1   Sperm competition

The realization that a female bird might mate with more than one male during a reproductive cycle has resulted in a dramatic shift in the way in which ornithologists perceive competition between males to pass on their genes. Prior to this paradigm shift, precopulatory competitive behaviour was presumed to be the means by which male precedence was determined. However, we now know that sperm competition, a postcopulatory phenomenon, is widespread and highly significant.

We saw in Chapter 4 that at copulation sperm are transferred from the male to the female and that they must then travel through the oviduct to the infundibulum in order to fertilize the egg. Not all of the transferred sperm make this journey. Some of them enter sperm storage tubules at the junction of the uterus and vagina (see Figure 4.1) where they can remain viable for periods of many days. Sperm from these tubules can be released to fertilize eggs produced over several days without the need for further copulation. Tim Birkhead and his coworkers at the University of Sheffield have established that

in the case of the Zebra Finch *Taeniopygia guttata* around 10% of eggs laid 13 days after the last copulation have been fertilized (Figure 5.1).

Birkhead and his colleagues have also investigated the effect of multiple male matings upon the paternity of eggs laid by a female. Their work has demonstrated very clearly the existence of within-female competition between the sperm of different males and the potential consequences of extra-pair copulations (EPCs). The results of some of these experiments are summarized in Figure 5.2.

The first experiment (Figure 5.2A) was designed to simulate a situation in which a single female mates with two males, switching rapidly between them but mating with each in turn. In this case, the result was that the second male (and last to mate with the female) fathered the majority of the eggs that were laid. The second experiment (Figure 5.2B) simulates the situation in which a female that mates regularly with her mate (the first male) is involved in just one EPC with the second male (significantly this EPC is the last copulation in the mating sequence). Again the

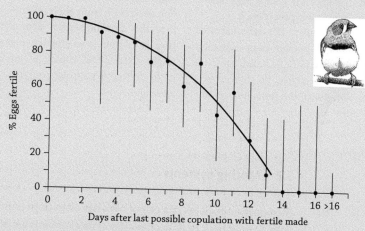

Figure 5.1 Egg fertilization levels decline with time following a copulation event but stored sperm may remain viable for up to 13 days. From Birkhead, T.R. and Møller, A.P. (1992) *Sperm Competition in Birds: Evolutionary Causes and Consequences*. Academic Press. Data from Birkhead, T.R., Pellat, J.E., and Hunter, F.M. (1988) Extra-pair copulation and sperm competition in the zebra finch. *Nature* **334**, 60–62.

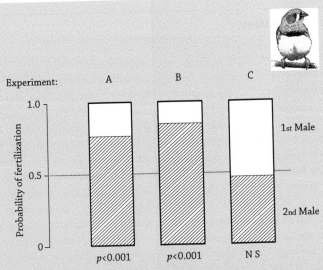

Figure 5.2 The outcomes of experiments (A–C) to investigate sperm competition in the Zebra Finch expressed as the proportion of eggs fertilized by each of two competing males (see text for explanations of individual experiments). From Birkhead, T.R. and Møller, A.P. (1992) *Sperm Competition in Birds: Evolutionary Causes and Consequences.* Academic Press. Data from Birkhead, T.R., Hunter, F.M., and Pellatt, J.E. (1989) Sperm competition in the Zebra Finch *Taeniopygia guttata. Animal Behaviour* **38**, 935–950.

figure shows that even this single mating by the second male results in his fathering the majority of the eggs that are laid. Taken together, these two experiments demonstrate sperm competition and a significant feature of this system—last male precedence.

The third experiment (Figure 5.2C) demonstrates the effectiveness of a well-known male strategy presumed to reduce the impact of EPCs and last male precedence—retaliatory copulation. In this experiment, each male copulates with the female once, but these copulations follow one another almost immediately (simulating the situation in which a male might observe his mate copulating with a rival or might infer a recent copulation having witnessed his female return from the territory of a rival). In this case the data presented demonstrate that retaliatory copulation is an effective means by which a male might reduce the impact of an EPC because both males fathered a similar proportion of the eggs that were laid.

5.2   Mating systems

The fidelity unto death of a pair of birds has often been presumed as being a given truth. On the basis of this commitment, pairs of birds are often used as a symbol of fidelity in human society (St Valetines day has bird associations for example). However, the advent of genetic paternity analysis has revealed that in fact socially monogamous birds are often quite promiscuous and a brood of apparent full siblings might in fact be sired by several males. Why should this be? Remember that

Table 5.1 Avian mating systems

System	Main features of system
Social monogamy	One male and one female cooperate to raise a brood of young. Genetic monogamy describes the situation in which these birds are both the genetic parents of the chicks being raised.
Polygyny	A male bird sires the offspring of a female and then deserts her (temporarily or permanently) to seek other females with which to mate. Deserted females raise young alone or with reduced help from a returning mate.
Polyandry	A female bird abandons her eggs to be raised by her male partner (this male is not always the genetic parent of the brood that he will raise). Deserting females may seek to mate with further males in the same breeding season.
Polygynandry	A group of males and females cooperate to rear young. The resulting broods are produced by several females and sired by several males.

male birds can maximize their reproductive output by fertilizing as many eggs as possible, and during that period of time when a female is incubating her eggs her male partner will often seek matings with other females (termed EPCs). Similarly, not all males are of equal genetic quality and so females paired to poor males will seek EPCs that will secure the genes of better males for their offspring.

Although some species of socially monogamous (one male and one female) birds may be genetically monogamous (i.e. no EPCs), more than 85% of those species that have been subjected to DNA paternity studies have been found to be sexually polygamous (multiple males and or females contributing genetic material to a single brood). So when thinking about mating systems it is necessary to consider both the social basis of the relationship between parent birds (do they work together to raise young or does one partner desert the other for example) and the genetic relationships between offspring and social parents (i.e. the birds which rear them but are not necessarily their biological parents). Table 5.1 provides a brief description of the main avian mating systems, many of which are described in more detail later in this chapter.

Box 5.2 Leks

A lek site is the traditional location at which a group of males come together to compete and display to visiting females. As a system, lekking, as the behaviour is termed, is rare, but it has evolved several times and is found in a number of bird groups.

Lekking is an unusual mating system in that males are chosen by the females as mates but they do not then provide any parental care or any direct territorial/resource benefit. The females gain nothing other than the genes that their offspring will inherit. One of the features of a lek is that matings are actually achieved by relatively few (sometimes only one) of the males that are present. This fact has been viewed as something of a paradox, the question asked being that if most males will not mate why do they attend the lek. In an attempt to resolve this paradox, four main hypotheses have been proposed; *hot-spots*, *hot-shots*, *kin selection*, and *female preference*.

The *hot-spot* hypothesis suggests that leks of males form in particular locations that are regularly visited by the females in a population. This would maximize the chances that males would encounter potential mates. Some support for this hypothesis does come from the observation that males typically choose particular habitat features when establishing a display arena. Male Houbara Bustard *Chlamydotis undulata* choose to lek in open areas where their displays can bee seen by females who more typically inhabit dry wadis and scrubby cover. On the other hand, the locations of the leks of Blue-crowned Manakin *Lepidothrix coronata* have been shown to be at sites no more likely to be visited than any other areas of their range. Similarly, when the dominant Great Snipe *Gallinago media* is removed from its position at the centre of a lek its place is not occupied by another bird; instead the lek collapses. This suggests that in this case hot-shots rather than hot-spots are significant. The *hot-shot* hypothesis suggests that inferior males cluster around a superior male hoping to gain access to at least some of the females attracted to him, and in some species it has been shown that those males with a territory/lek position close to the alpha bird do have better breeding success than those on the fringes of the lek.

The *kin selection* hypothesis suggests that leks are composed of related males in effect cooperating to attract females and gaining an indirect fitness benefit when genes that they have in common with a more successful relative are inherited by his offspring. It has been shown that captive male Peacock *Pavo cristatus* reared in isolation tend to form leks with relatives, but lek mates have been shown to be unrelated in a number of other species, including several of the manakins (Pipidae) and the Greater Sage Grouse *Centrocercus urophasionus*.

As an alternative, the *female preference* hypothesis suggests that leks form because females seek out larger groups of males in order to compare them and secure high quality males. If this were the case then we would expect to see a female preference for larger rather than smaller leks, and exactly this result has been obtained through observation of Little Bustard *Tetrax tetrax* lek sites.

Interestingly it has been suggested that leks might not be as far removed from conventional territorial breeding systems as was previously thought, and it has been suggested that clusters of territories may act as 'hidden leks' clustered around a high quality male or location (hot-shot or hot-spot), clustered around a group of related males (kin selection), or to facilitate female comparison of males (female preference).

Further reading

Fletcher, R.J. and Miller, C.W. (2006) On the evolution of hidden leks and the implications for reproductive and habitat selection behaviours. *Animal Behaviour* **71**, 1247–1251.

Högland, J. and Alato, R.V. (1995) *Leks*. Princeton University Press, Princeton.

5.3 Courtship and mate choice

As the 'rarer' sex, females are most often the choosier sex (but see Box 5.2); but on what basis are they making their choice? What do they look for in a mate? Observations suggest that some females choose on the basis of the quality of a resource provided to them by potential suitors. Arctic Tern *Sterna hirundo* females, for example, are more likely to pair with and be faithful to a mate who provides her with nuptial gifts of good quality fish during courtship (Plate 19). Males (and in some cases females) of some species establish a breeding territory, an area defended against conspecifics in which all of the resources needed for the successful rearing of young can be obtained. Ketterson and Nolan noted that those male Dark-eyed Juncos which return to their breeding grounds fastest and fight most vigorously to establish their territory first are also the ones that are chosen preferentially by females and have the highest productivity, indicating that having a good territory is a prerequisite to being a successful mate in many species.

Females of other species make the choice on the basis of the promise of a resource communicated by some signal (a song or display for example); females of some species seems to be making the choice on the basis of the signal itself rather than on the basis of the physical resource that it indicates.

Flight path: migrant male Dark-eyed Juncos take risks to establish the best territories. Chapter 3.

Resource provision

Great Grey Shrikes *Lanius excubitor* (Plate 20) are socially monogamous raptor-like passerines which form territorial breeding pairs. It is usual for both male and female to contribute to the raising of their young. Male shrikes are well known for their habit of impaling food items (large insects, small mammals, birds, and reptiles) on thorns or barbed wire fences, a habit that has earned them the title butcher birds. The quality of this larder (the size and nutritional value of the food it contains) has been shown experimentally to be one of the key factors used by a female when she chooses her mate. Larger prey items are presumed to require a greater effort on the part of the male in terms of his ability to capture them and the energy that he expends manipulating them; they are therefore likely to be an indicator of his physical quality.

I said earlier that these shrikes are socially monogamous. Remember that in saying this I mean that two birds (one of each sex) form a breeding pair which rears young together. You should also remember, however, that this does not imply that these birds are necessarily faithful to one another. Great Grey Shrike males do cuckold their neighbours, fertilizing an egg that is raised by an unsuspecting foster parent. But why do the females take part in these promiscuous liaisons; surely they have a suitable mate? Piotr Tryjanowski and Martin Hromada have made detailed observations of a population of these birds and have shown that in fact the males effectively pay for sex with already mated females by providing them with choice items of food from the larders that initially impressed their own mates.

Their data (Figure 5.3) show that males are more likely to be successful when they offer a female a better gift, and that they tend to offer the best gifts (the most energetically valuable) to their 'mistresses'. These gifts they estimate to contribute 66% of the daily food requirement of a female (compared with the 16% they provide their mate). Cheating females benefit from these EPCs because they have an opportunity to mate with high quality males (presumably males of higher quality than their own mates). Their cuckolded partners, however, do lose out because they will invest resources raising the chicks of a rival male. However, it would be quite wrong

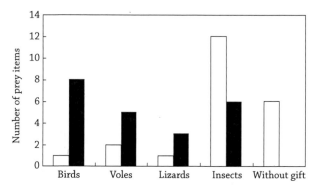

Figure 5.3 Male Great Grey Shrikes provide more (and better) gifts when seeking extra-pair copulations (solid bars) than when seeking to mate with their partner (open bars). Tryjanowski, P. and Hromada, M. (2004) Do males of the great grey shrike *Lanius excubitor*, trade food for extrapair copulations? *Animal Behaviour* **69**, 529–533.

to think of these hapless males as being unable to avoid the costs associated with promiscuity and there is abundant evidence that males do their utmost to avoid being the 'victims' of EPCs.

Francisco Valera and coworkers have made similarly detailed observations of the breeding behaviour of a population of Lesser Grey Shrikes *Lanius minor*, a close relative of the Great Grey and another butcher bird that uses food as a nuptial gift. They noted that territorial intrusions by males that they presumed to be actively seeking EPCs were seven times more common during the fertile period of the resident female. During this period the resident male was particularly attentive to his mate, spending almost 80% of his time within 50 m of her. He was also particularly aggressive towards male intruders, attacking them and chasing them away. So mate guarding seems to be an effective EPC minimization strategy. But what about the 20% of the time he didn't guard his mate? To find out what would happen if males had apparent reason to suspect that an EPC had occurred, the researchers captured females during their fertile period and removed them from their territory for 1 hour. They then released them back into an adjacent territory so that when they returned to their mate it would appear that they had visited his rival. Males responded to this apparent infidelity by punishing their mate, attacking her, and in many cases aggressively forcing copulation. Such retaliatory copulations are of course an important paternity assurance strategy (see Box 5.1). The same did not, however, happen if the removal had taken place outside of the fertile period (during incubation or chick rearing for example). There is also a suggestion that unlike males the females in this population are reluctant to seek EPCs; they rarely leave their territory during their fertile phase even if their male has been temporarily removed. Genetic paternity analysis has revealed that male mate guarding and female punishment are an effective strategy in the case of this species because mixed paternity broods are extremely rare.

Key reference

Valera, F., Hoi, H., and Krištín, A. (2003) Male shrikes punish unfaithful females. *Behavioural Ecology* **14**, 403–408.

Ornaments and displays

Male Long-tailed Widowbirds *Euplectes progne* have, as their name suggests, an extremely long tail. In fact this 15–20 cm long bird can have a tail around 50 cm long during the breeding season. We saw in Chapter 2 that feathers can be expensive, and there must surely be an aerodynamic cost to be paid when towing a streamer this long. So why do these males have such a long tail? It seems to have evolved as a result of sexual selection because there is good experimental evidence of a link between tail length, male reproductive success, and female choice. Simply put, females want males with long tails. The evidence for this comes from a particularly elegant experiment carried out by ornithologist Malte Andersson who captured males and artificially elongated or shortened their tails and then observed the effect that this had upon their subsequent reproductive success. Long-tailed widowbirds are polygamous. Males defend an area of grassland from other males and advertise their presence to females by means of a conspicuous bouncing display flight showing off their tails to good effect. Each male attempts to attract a harem of females to his territory, all of whom will rear his young (assuming no EPCs take place of course) without his assistance. Andersson trapped male birds and assigned them randomly to three groups. One set of birds had their tail feathers cut in half and rejoined with no net change in length (this was done as an experimental control). Another group had a 25 cm long section of their tail removed and the tip was rejoined to the base—thereby shortening the tail. These sections were inserted into the cut tails of the final experimental group, thereby lengthening their tails. Andersson then released them back into their territories and recorded the number of additional nests that each bird built (a measure of the number of additional females attracted to him). The results (Figure 5.4A) show clearly that males with the longest tails attract the most mates.

However, the results of another tail lengthening experiment carried out by Sarah Pryke and Staffan Andersson (Figure 5.4B) provide an intriguing insight into the basis of the evolution of this courtship signal. They manipulated the tails of a relatively short-tailed species, the Red-shouldered Widowbird *Euplectes axillaries*. Males of this species are slightly smaller than male Long-tailed Widowbirds but they have very much shorter tails (around 7 cm long). When these tails were artificially lengthened (in some cases to 22 cm long) males attracted as many as six females; three times as many as the longest tailed unmanipulated individual. So it seems that female widowbirds have a generalized preference for longer tails, and that this sensory bias has driven the evolution of the extraordinarily long tails of some species.

Sharing a mate

Female widowbirds seem content to share their male with others in his harem. This is probably because all they require of him are his genes and access to the resources available in his territory. They rear their young without his assistance.

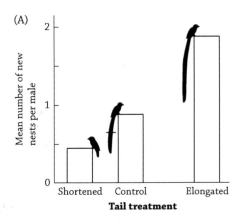

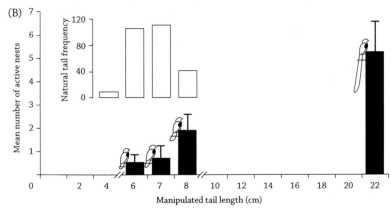

Figure 5.4 The relationship between male tail length and reproductive success in the Long-tailed Widowbird (A) and the naturally short-tailed Red-shouldered Widowbird (B). From (A) Andersson, M. (1982) Female choice selects for extreme tail length in a widowbird. *Nature* 299, 818–820; (B) Pryke, S.R. and Andersson, S. (2002) A generalized female bias for long tails in a short-tailed widow bird. *Proceedings of the Royal Society B: Biological Sciences* **269**, 2141–2146.

Box 5.3   Turning the tables: reproductive role reversal in the Spotted Sandpiper

In the vast majority of bird species males compete with one another for the attentions of choosier females, and females are able to make the most of good males and favourable breeding conditions by varying the numbers of eggs that are laid in a clutch. The Spotted Sandpiper *Actitis macularis* is an interesting exception to this rule. Female Spotted Sandpipers always lay four eggs in a clutch. Unable to vary their clutch size, the only way that they could increase their output in a good year would therefore be to lay a second clutch of four eggs. However, the sandpiper breeding season is a short one and it is doubtful that a female could manage to rear one brood of chicks to fledging and then have time to rear a second before migrating.

Faced with these difficulties female Spotted Sandpipers employ an interesting strategy—they act like males. Females arrive on the summer breeding grounds first and compete with one another to secure territories. When the males do arrive the females actively court them and the 'best' females secure mates quickly and lay a clutch of four eggs.

When resources allow it the females of some populations then abandon their newly laid eggs, leaving their mate to incubate and rear the brood alone. Males are only able to do this because the sandpipers time their breeding to coincide with the emergence of a superabundance of insect food and because newly hatched sandpipers are precocial (able to thermoregulate and fend for themselves from hatching). Furthermore, the system works because males are slightly more common in the population than females and so after abandoning their first mate females are able to secure a second and lay a second clutch of eggs to be reared by him.

The first males to pair up clearly benefit because they have secured for their offspring the genes of the best females and because the best that they can do is secure for themselves a clutch of four eggs. But do the females compromise the quality of their offspring when they take a second male? After all, these are the males that were 'left on the shelf' first time around and so are presumably inferior in some way. In fact, DNA paternity analysis has revealed that second brood chicks are often fertilized by first male sperm that have been stored by the female; in this way females are able to make the most of the genetic resources available to them.

Reference

Oring, L.W. *et al.* (1992) Cuckoldry through stored sperm in the sequentially polyandrous spotted sandpiper. *Nature* **359**, 631–633.

In some species, however, polygyny appears to result in reduced reproductive success for at least some of the females involved. So why do they accept it?

In 1969 Gordon Orians published a very influential paper in which he proposed a mathematical model to explain female acceptance of polygyny. In his polygyny threshold model (PTM), Orians envisaged a situation in which females would sample and compare the territories of available males and then, using the information that they have gathered, elect to join an already mated male in a polygamous relationship or to settle with an unmated male in a monogamous one. This model assumes that females should prefer monogamy because there will be a cost to polygamy and that females should only accept polygamy when the benefits of that relationship outweigh its costs relative to monogamy with an available unmated male. The point at which this economic decision is made is the polygamy threshold (see Figure 5.5).

Do female birds really behave in a manner consistent with the PTM? Stanislav Pribil and William Searcy have demonstrated experimentally that in the case of at least one species, the Red-winged Blackbird *Aegaius phoecniceus*, they do. The PTM makes a number of testable assumptions: (1) that polygamy is costly to females, so they prefer monogamy; (2) that females choose males on the basis of either male quality or the quality of the territory of the male; and (3) that faced with a choice between a low quality monogamous male/territory and a high quality polygamous male/territory, females will choose polygamy (if the cost of polygamy is lower than the cost of choosing monogamy in this situation). Through their observations and experiments Pribil and Searcy have shown that

Key reference

Orians, G.H. (1960) On the evolution of mating systems in birds and mammals. *American Naturalist* **104**, 589–603.

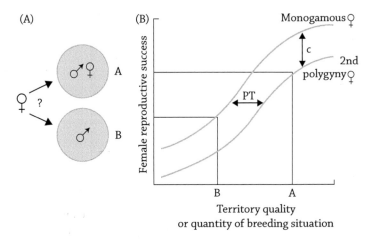

Figure 5.5 The polygyny threshold model. The model presumes that females have choice and can choose to pair with already mated or unmated males (A). Female reproductive success varies according to the quality of the territory of the male chosen. If a female can gain more by choosing an already mated male than she can by choosing a single bird then she should choose polygamy over monogamy (B). From Scott, G.W. (2005) *Essential Animal Behavior.* Blackwell Science, Oxford. Adapted from Orions, G.H. (1969) On the evolution of mating systems in birds and mammals. *American Naturalist* **104**, 589–603.

Key reference

Pribil, S. (2000) Experimental evidence for the cost of polygyny in the red-winged blackbird *Agelaius phoeniceus. Behaviour* **137**, 1153–1173.
Pribil, S. and Seacrcy, W.A. (2001) Experimental confirmation of the polygyny threshold model for red-winged blackbirds. *Proceedings of the Royal Society B: Biological Sciences* **268**, 1643–1646.

all three of these assumptions are met in the case of their Red-winged Blackbird study populations.

Pribil began by confirming the first prediction of the model when he demonstrated that female Red-winged Blackbirds suffered a reproductive cost if they chose to mate with an already mated male; he found that they fledged fewer and lighter young. This observation is important because survival to maturity in this and many species is significantly correlated with fledging weight; heavier birds survive better. As would be predicted from these results, he also showed that when offered the choice of two males, one with a mate and one without, experimental females invariably chose to enter into a monogamous relationship.

When they have the choice, female Red-winged Blackbirds also discriminate between males on the basis of territory quality (the second of the model's predictions). They will preferentially mate with an unmated male controlling a territory that offers the opportunity to build a nest overhanging water (presumably because such nests offer enhanced protection from predators).

To test the model's third and perhaps most significant prediction—that faced with a choice between a low quality monogamous male/territory and a high quality polygamous male/territory, females will choose polygamy (if the cost of polygamy is lower than the cost of choosing monogamy in this situation)—Pribil and Searcy designed an elegant field-based experiment in which they manipulated the choices available to females. In it they compared the attractiveness to females of males whose territories were manipulated such that one in each pair offered an unmated male with no overwater nest site while the other offered a high quality nest (over-water, see Plate 21) that was occupied by an already mated male. In almost all cases newly arriving females chose polygyny rather than monogamy, exactly as would be predicted by the PTM.

Box 5.4   The dunnock: a case study in sexual conflict

Within a single population of the Dunnock *Prunella modularis* it is possible to recognize genetic monogamy, social monogamy, polygyny, polyandry, and even polygynandry. As a consequence, the study of the reproductive behaviour of this otherwise unobtrusive bird by Nick Davies and his Cambridge University colleagues is quite possibly one of the best, and best known case studies of bird breeding behaviour.

Central to the system is sexual conflict, i.e. the conflict of interests of males and females. Figure 5.6

summarizes the benefits and costs of each system (in terms of chicks produced) to both males and females.

Taking monogamy as a starting point, males and females both benefit equally—raising 5 young each (assuming that this is genetic and social monogamy). But note that a female can improve upon this situation if she persuades a secondary male to join a polyandrous group—she will raise 6.7 chicks, and the secondary male benefits because rather than have

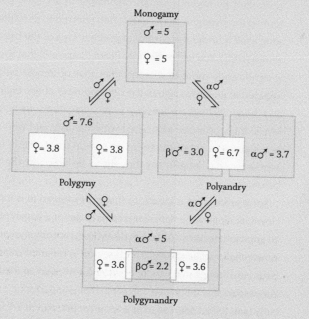

Figure 5.6 The complex mating system of the dunnock. Several scenarios are illustrated, and in each case the average number of young produced per individual is given. See text for detailed explanation. From Davies, N.B. (1992) *Dunnock Behaviour and Social Evolution.* Oxford University Press, Oxford.

none he will have sired 3 chicks. Female dunnocks regularly court males in an attempt to benefit in this way. However, this arrangement clearly does not benefit the primary male; rather than siring 5 chicks he sires just 3.7 on average. Primary males therefore attempt to drive away secondary males.

A similar state of affairs exists in the case of males in that they can do considerably better if they persuade a second female to join them in a polgynous group, and if polyandry is inevitable then by attracting another female to change a polyandrous group into a polygynandrous one he can at least not lose out.

The actual system that is established will represent a compromise on the part of one or all of the parties involved, and of course through EPCs and EPC countermeasures the actual genetic composition of broods may be even more complicated than Figure 5.6 would suggest.

5.4 Bird Song

Male birds of a number of species sing a courtship song to attract a mate and as a form of resource defence (Plate 22). We will consider both of these function of song below, but first we should consider the song and singing behaviour itself. Bird song is produced when air passes through the syrinx, the avian equivalent of the human voice box. Variations in song are a result of the carefully controlled contraction of the muscles and membranes of the syrinx which are, in turn, controlled by nervous signals originating in a clearly defined area of the brain termed the hypoglossal nucleus (often referred to as nXIIts). But the production of song itself is under the control of another area of the brain termed the high vocal centre (HVC). You will remember (from Chapter 4) that the HVC of male songbirds is significantly more developed than that of female birds and this difference is initiated at an early stage in the development of the brain. Cells in both the HVC and the syrinx itself are sensitive to the male hormone testosterone, and increased levels of testosterone associated with maturity and with the development of the testes at the onset of the breeding season (triggered in temperate regions by an increase in day length) are thought to be amongst the triggers of male singing behaviour.

Song production is innate, in that males reared in isolation will sing at maturity. However, singing the right song seems to depend largely upon a male learning from an appropriate tutor. (Although it has been recently shown that in the Reed Warbler *Acrocephalus schoenobaenus* birds reared in isolation can mature to sing normally.) Birds learn songs in a variety of ways. Some such as the White-crowned Sparrow *Zonotrichia leucophrys* can only learn their songs during a very short period of their early lives (often referred to as a sensitive period). In the Chaffinch *Fringilla coelebs* this sensitive period is longer (lasting for the whole of the first year of a bird's life). Such birds are termed closed-ended learners or age-limited learners. On the other hand, males of some species such as the Willow Warbler *Phylloscopus trochilus* are able to learn new songs and add to their repertoires throughout their lives.

Flight path: development and control of bird song. Chapter 4.

Key reference

Chen, X., Agate, R.J., Itoh, J., and Arnold, A.P. (2005) Sexually dimorphic expression of trkB, a Z-linked gene, in early posthatch zebra finch brain. *Proceedings of the National Academy of Sciences, USA* **102**, 7730–7735.

Song learning

The various mechanisms by which song learning takes place are proving to be more variable than was originally thought to be the case, and I would recommend that the interested reader consult the excellent review of the topic by Beecher and Brenowitz. However, the basic model of learning seems to include three steps: a preliminary sensory acquisition phase; a silent phase; and, finally, a sensorimotor phase.

Young birds seem to have an innate template that can be modified to some degree by a process of learning. During the first phase of song development (the sensory acquisition phase which equates to the sensitive period) a young bird will hear a wide range of noises, but evidence suggests that they are most sensitive to the song that most closely matches their own innate template—the song of a suitable conspecific tutor. In some cases such as the Zebra Finch *Taeniopygia guttatta* and Galapagos finches (Geospizidae) the tutors are always, or almost always, the rearing male parent. In others, however, the song that a bird learns is not a complete copy of that of its father. In a study of a wild population of Savannah Sparrow *Passerculus sandwichensis*, Nathaniel Wheelwright and colleagues have shown that males in their study population developed a song similar to (but not the same as) that of their social father in only 12% of cases, the majority of the population learning a song more similar to that of their neighbours. Their results also suggest that in this population individuals incorporate elements of song from a number of tutors into their repertoire.

During this learning phase it appears that a bird memorizes aspects of tutor song and uses them to refine its original innate template, thereby producing a more exact one that conforms to the specific song pattern of its species. Some birds in isolation will learn from a recording, but learning does appear to be enhanced when a live tutor is present and, in some cases, such as the Zebra Finch *Taeniopygia guttatta*, a live tutor is essential. After song acquisition there follows a silent phase when singing is not taking place but during which the components of song that the bird has learned are stored for future use. Prior to the onset of the breeding season (when the testes regenerate) testosterone triggers singing, and the final phase of song acquisition, the sensorimotor phase, occurs. Now the bird sings. Initially the song that is produced will not be perfect but it will be a very close approximation of the song typical of the species. Through time (and actually very quickly) the song is refined until it is perfected or crystallized. It seems that hearing oneself is crucial at this time and that birds match the songs that they produce against their now completed internal template. In a study of chaffinches, Nottebohm found that birds artificially deafened early in the silent phase lacked the ability to perfect their song whereas birds deafened as adults were able to sing normally.

Functions of song

Bird song has two primary functions. It is used by male birds both as a courtship signal to attract mates and as a signal to other males that a territory is occupied

Key reference

Beecher, M.D. and Brenowitz, E.A. (2005) Functional aspects of song learning in songbirds. *Trends in Ecology and Evolution* **20**, 143–149.

Key reference

Wheelwright, N., Swett, M.B., Levin, I.I., Kroodsma, D.E., Freeman-Gallant, C.R. & Williams, H. (2008) The influence of different tutor types on song learning in a natural bird population. *Animal Behaviour* **75**, 1479–1493.

Key reference

Nottebohm, F. (1968) Auditory experience and song development in the chaffinch *Fringilla coelebs*. *Ibis* **110**, 549–568.

(see Box 5.5). Artificially muted birds have been shown to be unable to either gain territories or attract mates in a number of studies.

The temporal coincidence of male singing behaviour and the onset of reproductive behaviour is in itself highly suggestive that singing has a courtship function, and evidence that this is the case comes from a range of field and laboratory studies. Singing rates of territorial males have been shown to be higher before a female joins a male on his territory than they are once the pair is established. However, if a female is temporarily removed from the territory the resident male responds by increasing his singing rate. Presumably he is attempting to replace his 'lost' mate. Similarly males may increase their singing behaviour while their mate incubates their clutch of eggs. Presumably these birds are attempting to attract a second mate or to secure EPCs. Dag Eriksson and Lars Wallin have demonstrated very clearly that male song attracts females in the case of the Collared Flycatcher *Ficedula hypoleuca* and the Pied Flycatcher *Ficedula albicollis*. These birds nest readily in nest boxes, and the boxes can be used to trap birds that enter them (trapped birds are released unharmed very quickly). The researchers arranged 28 nest boxes throughout their study population, each box having a model flycatcher about 1 m from it on a prominent perch (live flycatcher males usually sing from such perches presumably to attract females to them). From half of the boxes they played the song of the same species as the male model perched outside. The other half of the boxes were silent. The results of this study clearly show that females were far more likely to inspect the boxes of singing males, and song can therefore be presumed to have attracted them to the territory and persuaded them that the next box might make a suitable home. About 90% of the female flycatchers trapped whilst inspecting nest boxes were attracted to those boxes having both a model male and a recording of his song.

Key reference

Eriksson, D. and Wallin, L. (1986) Male bird song attracts females—a field experiment. *Behavioural Ecology and Sociobiology* **19**, 297–299.

Box 5.5   What's in a song?

European Starlings *Sturnus vulgaris* have been described as being amongst the most accomplished of singers. Males have extremely varied repertoires and sing for prolonged periods during their breeding season. To demonstrate that their song has both a female attractant and male deterrent function, James Mountjoy and Robert Lemon studied a wild population of starlings (breeding readily in nest boxes). To see if the birds would be attracted to it, starling song was played from some of the boxes. As a control, and for comparison, each box was paired with a second, from which no song was broadcast. Throughout their observation period the team saw no female starlings at silent boxes, but they saw 12 investigating the boxes from which song was being broadcast. The song was evidently attractive to them. We might have expected male starlings to behave differently—given that we assume song to serve as a deterrent to rivals—but in fact of 20 birds seen at boxes, 17 were investigating the ones with a playback. So is song a deterrent to rival males in this species? A further refinement of the experiment showed that it is.

In a second experiment they paired boxes with a playback of a very simple starling song with boxes playing a very complex one. This time the sexes did behave differently. All of the females observed were attracted to the complex song whereas 90% of the males were attracted to the simple song. So it would seem that a complex song would benefit a male because it would attract females and deter male rivals.

Further evidence that female starlings consider males with a complex song to be higher quality comes from another piece of research carried out by Mountjoy and Lemon. They made further detailed observations of their study population, this time recording the complexity of the song of individual males during the courtship period and then noting the date at which the first egg was laid in the boxes occupied by each of these males. Their prediction was that the males with the most complex songs would be in the best condition and would therefore be the most attractive to the females. The females should therefore be prepared to commit to pair with and lay eggs with these males preferentially.

From Figure 5.7A it is clear that a positive relationship between song complexity and body condition (in this case an expression of size and mass)

exists—fitter birds are better singers. As would be predicted Figure 5.7B shows that females paired to males with the biggest repertoires lay eggs sooner than those that pair to poorer males.

Recently research has revealed another potentially important indication that song could be used by female starlings to assess the quality of potential mates (and presumably, therefore, to influence the potential quality of their eventual offspring). Deborah Duffy and Gregory Ball have demonstrated a correlation between the male song variables bout length (the mean length of a single song) and singing rate (the mean number of times a bird sings in an hour) and two measures of the strength of the immune system of the birds. In both cases their results demonstrated clearly that better singers had the most robust immune system.

References

Duffy, L.D. and Ball, G.F. (2001) Song predicts immunocompetence in male European starlings (*Sturnus vulgaris*). *Proceedings of the Royal Society B: Biological Sciences* **269**, 847–852.

Mountjoy, D.J. and Lemon, R.E. (1991) Song as an attractant for male and female European Starling, and the influence on song complexity on their response. *Behavioural Ecology and Sociobiology* **28**, 97–100.

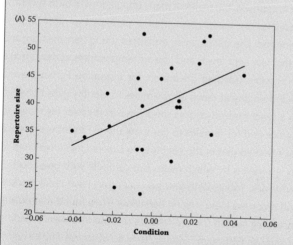

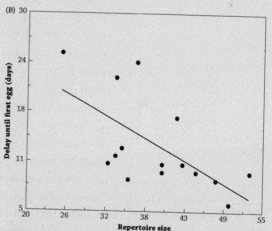

Figure 5.7 (A) The relationship between male starling song repertoire size and body condition of male. (B) The relationship between male starling song repertoire size and delay until the laying of the first egg. From Mountjoy, D.J. and Lemon, R.E. (1996) Female choice for complex song in the European Starling: a field experiment. *Behavioural Ecology and Sociobiology* **38**, 65–71.

Females are able to discriminate between males on the basis of their songs, preferring some song types over others. Female dunnocks, for example, have been shown to pay more attention to a recording of the song of their mate when he has been removed than to the song of a neighbour. Furthermore, this is particularly the case during the female's fertile period, suggesting that she uses her mate's song as a way of finding him to seek out copulations.

Synchronized singing

Although the songs of individuals of a species do vary, often markedly, a consequence of birds learning from those around them may be that all of the birds in an area develop a broadly similar repertoire. This seems to be an advantage in some cases, and males are known to match their repertoires during singing contests. It appears that matching your song repertoire to that of a rival can reveal to him that you are a familiar neighbour and may therefore be tolerated. On the other hand, a song repertoire that is different is more likely to be responded to aggressively because it reveals the singer to be an unknown intruder and therefore to be a greater potential threat.

Song matching, however, i.e. the singing of the same song as a rival at the same time as a rival, and the overlapping of singing bouts between rivals do appear to be particularly aggressive signals. Birds matching or overlapping in this way are more likely to escalate their contest to a full-blown fight. Perhaps not surprisingly these contests are more common between strangers or between birds establishing territories at the start of a breeding season than between established territorial neighbours.

Perhaps to facilitate comparison, the males of many populations of birds synchronize their singing behaviour, the dawn chorus being perhaps the most familiar example of this phenomenon. The still dawn air enhances sound transmission, and it has been suggested that low light levels and low air temperatures at dawn make other behaviours less possible (feeding on insects for example). Perhaps overnight mortality is high and singing at dawn allows males to identify gaps between territories. Or perhaps a synchronized chorus simply makes it easier for males to compete with one another and for females to compare them. Recently, evidence from radio tracking has revealed that in the Nightingale *Luscinia megarhynchos*, at least, the dawn chorus is a means by which males can compete with one another and identify vacant territories. Tobias Roth and his colleagues have radio tracked female nightingales and recorded male singing behaviour. They found that during the early part of the breeding season (before females migrate into their breeding areas) males sing most at dawn (although famously they do of course sing all night) (Figure 5.8A). Once the females arrive, paired males continue to be most vocal towards the end of the night and at dawn (Figure 5.8B), but bachelor males increase their singing throughout the night. Radio-tracked females (released into the area by the researchers to simulate new arrivals) were found to be most mobile at night

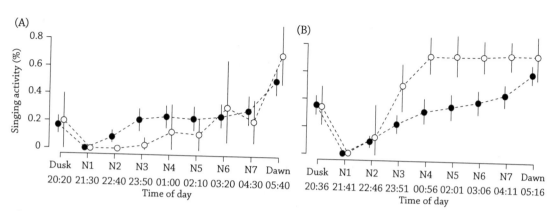

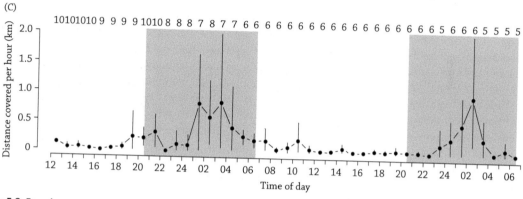

Figure 5.8 Female movement and male singing behaviour of nightingales. (A) Prior to female arrival, birds that will eventually find a mate (closed circles) and those that will not (open circles) both increase their singing activity as dawn approaches. (B) After female arrival, paired birds and birds that will eventually find a mate (filled circles) sing most at dusk and as dawn approaches. (C) Females searching for prospective mates are most active at night. From Roth, T., Sprau, P., Scmidt, R., Naguib, M., and Valenitin, A. (2009) Sex specific timing of mate searching and territory prospecting in the nightingale: nocturnal life of females. *Proceedings of the Royal Society B: Biological Sciences* **276**, 2045–2050.

(Figure 5.8C). Roth and his coworkers interpret these observations as follows. They suggest that newly arrived females visit several males over the course of a night, listening to the song of each of them prior to choosing a mate (this is why bachelors rather than paired males sing at this time). At dawn the females become inactive and the function of singing switches from mate attraction to territorial defence (all male territory holders sing at this time).

A further advantage of synchronized singing may be that it helps the birds of an area to synchronize the rest of their reproductive behaviour. Doing so could be an advantage in that synchronized production of young may swamp local predator populations with prey and thereby ensure survival of a greater proportion of young birds and increase the individual survival probability of each chick. It may also enable sharing of nest/chick defence activity and improve the efficiency of foraging parents (if they are able to feed as flock mates with other foraging parents). In the case of Zebra Finch *Taeniopygia guttatta* Joseph Wass and his colleagues have shown

> **Flight path:** swamping predators can reduce individual vulnerability to predation. Chapter 6.

Box 5.6 Bird song and noise pollution

In 2006 Hans Slabbekoorn and Ardie den Boer-Visser coauthored a report with the attention-grabbing title *Cities Change the Songs of Birds*. In this thought-provoking article they made the chilling claim that worldwide urbanization and the ongoing rise of urban noise levels form a major threat to living conditions in and around cities. Specifically they were highlighting the problem faced by songbirds having to compete with urban noise pollution to make themselves heard. This issue is not a new one; for some time researchers from around the world had demonstrated repeatedly that breeding bird densities are lower and bird communities are less diverse close to noisy motorways. Frank Rheindt, who censused the birds breeding along woodland transects moving away from a busy German motorway, showed that in general those species with higher pitched songs were less susceptible to this noise form of noise pollution. But he demonstrated that some species such as the Chiff-Chaff *Phylloscopus collybita* and Great Spotted Woodpecker *Dendrocopos major* with low frequency songs were 60–75% less common closer to the road. It seems likely that low frequency traffic noise masks low frequency bird song and hampers courtship and territorial defence.

In their study, Slabbekoorn and den Boer-Visser compared the songs of Great Tits *Parus major* breeding in ten major European cities with woodland populations close to each of them. Like Rheindt they also suggest that low frequency urban noise pollution masks low frequency bird song. But in this case they also found that by altering their songs the birds were fighting back. Analyses of the songs of the Great Tits in their study populations revealed a frequency shift in urban birds. They were no longer singing the lower notes of their song at a low frequency. Although the upper frequency of their song remained unchanged the lower notes were now higher in urban than rural settings and so the frequency range of their song was narrower. This ability to respond and adapt might seem like good news, and for the Great Tits it probably is, but, as Alejandro Rios-Chelen points out in his 2009 review of the topic, it is far from clear that all species can adapt in this way and we do not yet know the impact of the adaptation in terms of sexual selection and the ability of males to attract mates.

References

Rheindt, F.E. (2003) The impact of roads on birds: does song frequency play a role in determining susceptibility to noise pollution? *Journal für Ornithologie* **144**, 295–306.

Rios-Chelen, A.A. (2009) Bird song: the interplay between urban noise and sexual selection. *Oecologia Brasiliensis* **13**, 153–164.

Slabbekoorn, H. and den Boer-Visser, A. (2006) Cities change the songs of birds. *Current Biology* **16**, 2326–2331.

Key reference

Wass, J.R., Colgan, P.W., and Boag, P.T. (2005) Playback of colony sound alters the breeding schedule and clutch size in zebra fince (*Taeniopygia guttata*) colonies. *Proceedings of the Royal Society B: Biological Sciences* **272**, 383–388.

that exposure to the sounds of a breeding colony of finches (including male courtship song) caused male zebra finches to increase their own singing rate (particularly if the sounds were recorded from their own colony). They also found that females were more synchronous in their egg laying when they were played colony sounds, and they produced larger clutches of eggs.

5.5 Raising a family

You will recall from Chapter 4 that most young birds require some degree of parental care, the exception being the fully independent superprecocial chicks of some megapodes species. The degree of care required varies from a relatively low level of protection and tuition (precocial chicks) to that required by the altricial passerines that are blind, deaf, and completely helpless at hatching. During the posthatching period chicks are not efficient thermoregulators and rely to a large extent upon their parents for warmth or for shade (when the problem is an inability to stay cool). Both precocial and altricial chicks develop quickly though, and after about a week most are largely able to control their own body temperature. To do this they rely upon their insulating down and growing contour feathers to regulate heat loss, and upon the heat-generating shivering of their developed leg and breast muscles.

Rapid growth requires a lot of energy and so chicks typically have prodigious appetites. At this time birds often have a diet that differs from that of adults of their species to enable them to obtain higher than usual amounts of protein, fat, and calcium for muscle and bone development. Most passerine chicks, for example, are raised on a diet of soft-bodied insects, snails, and fragments of egg shell even if as an adult their diet would be restricted to grains and fruits. Some specialists such as pigeons and penguins regurgitate a nutritious mixture of fats and protein to facilitate very rapid chick growth. The 'milk' regurgitated by pigeons is composed largely of sloughed off oesophageal epithelial cells.

Begging

Chicks solicit food from their parents by begging, a behaviour which typically involves a screaming call, a wide gape, and exaggerated head movements. In the case of some birds, and particularly those in darker nests, the flanges of the gape and the palate are often brightly coloured to make them a more conspicuous stimulus (Plate 23). Initially most young birds are indiscriminate in their begging behaviour. For example, the very young chicks of cavity-nesting passerines will beg at any shadow passing in front of the nest entrance, be it friend or foe. As they mature their behaviour becomes more refined and eventually they become more discriminating; directing their begging only towards their parents and responding to other shadows with silence and a crouch.

In a set of experiments now quite properly regarded as being 'classics', Jack Hailman demonstrated this phenomenon in the case of the Herring Gull *Larus argentatus*. These chicks had been previously shown, by Niko Tinbergen, to peck instinctively at any beak-like stimulus just so long as it was vaguely similar to a real beak (long, thin, and with a contrasting mark towards its tip). Herring Gull beaks are yellow with a red spot towards the tip of the lower mandible. Chicks peck at the spot instinctively and the pecking stimulates the adult bird to regurgitate food—so pecking is a begging behaviour. Prior to Hailman's work it had been assumed that instinctive behaviours such as this one were inflexible, but by presenting wild Herring Gull chicks in their own nests with models of Herring Gull heads and beaks and Laughing Gull *Larus atricilla* heads and beaks (all red) he showed that this was not in fact the case. As the data presented in Figure 5.9 show, the chicks did initially

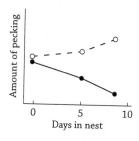

Figure 5.9 Although equally likely to beg from (peck) models of both Herring Gulls (open symbols, broken line) and Laughing Gulls (filled symbols, solid line) when newly hatched, Herring Gull chicks learn to focus their attention on a model of their parent. From Hailman, J.P. (1969) How an instinct is learned. *Scientific American* **221**, 106.

Key reference

Moreno-Rueda, G., Soler, M., Soler, J.J., Martinéz, J.G., and Pérez-Contreras (2007) Rules of food allocation between nestlings of the Black-billed Magpie *Pica pica*, a species showing brood reduction. *Ardea* **54**, 15–25.

Flight path: magpie broods suffer loses at the nestling stage. Chapter 4.

Key reference

Porkert, J. and Špinka, M. (2006) Begging in Common Redstart nestlings: scramble competition or signalling of need? *Ethology* **112**, 398–410.

peck at both stimuli, but through time their interest in the Laughing Gull model waned. Why? Well because these chicks were of course being fed by their parents and so learned to associate a Herring Gull head and beak with food and to ignore the inappropriate stimulus of a Laughing Gull head, a process termed perceptual sharpening.

Recently further subtleties of the begging relationship have been explored. Clearly chicks beg to be fed, and parents should respond to begging by feeding chicks; but do the needs of the parents and chicks and of siblings in the nest always coincide?

Gregorio Moreno-Rueda and his colleagues have demonstrated that Black-billed Magpie parents respond to the differing needs of their brood by feeding the chick that begs with the highest intensity. As begging intensity is known to correlate strongly with hunger in this and other species, this does seem to be a sensible strategy on the part of the adult birds. What would happen though if insufficient food was available and the whole brood could not be successfully fledged? (In most birds there is a strong relationship between size/mass at fledging and subsequent survival rates.) By always feeding the hungriest chick the parents may disadvantage the strongest and therefore reduce their own reproductive success. Moreno-Rueda's observations suggest that a further refinement of the behaviour of the parent birds allows them to avoid this problem.

Although the parent magpies do preferentially feed chicks that beg intensely they also preferentially feed larger chicks, and because magpie eggs hatch asynchronously there is often quite a size range amongst chicks within a brood. In a poor season, therefore, smaller chicks will in effect be allowed to starve so that their siblings can survive.

Brood reduction of this type is not an uncommon response to resource limitations, and, in many cases, such as some of the herons, raptors, and parasitic cuckoos, it is the norm. It is often more brutal, and competitor chicks are ejected from the nest, killed, and/or cannibalized by their siblings. However, it is not the case that all brood mates compete in this cut-throat way. For example, Jiři Porkert and Marek Špinka have shown that in the Common Redstart *Phoenicurus Phoenicurus* begging intensity is an honest signal of need (hungrier chicks beg more) and that like magpies adult redstarts do preferentially feed the hungriest chicks. In this species chicks within a brood vary very little in weight at fledging, suggesting that there is little competition between them. So do the parent redstarts have other feeding preferences? Yes they do. They preferentially feed those chicks that beg most closely to the entrance of the nest cavity. So does this mean that some chicks are fed more than others as was the case in the magpie? Through observation of the behaviour of the nestlings within broods, Porkert and Špinka found that once it had been fed to satiation a chick typically moved to the back of the nest, allowing its hungry nest mates to take their turn at the front.

This may not mean, however, that redstart parents feed all chicks equally. You may recall that in Chapter 4 I mentioned that female birds vary the investment that they make in each of the eggs that they lay. In effect they give favoured offspring a head start. We have also seen that this kind of differential parental investment persists after eggs have hatched, with some chicks receiving more food than others. It has been recently shown that this phenomenon can be more sophisticated than we might have thought. Male Spotless Starlings *Sturnus unicolor* have been shown to provide more food for chicks that hatch from darker shelled eggs. By doing this, the males are thought to favour the chicks that have hatched from the highest quality eggs. It is also the case that by some as yet undetermined means male birds are able to assess the likelihood that the chicks in their brood are unrelated to them, and therefore the result of EPCs. Having done so, they allocate their feeding accordingly and preferentially feed those chicks most likely to have been sired by them. Sometimes of course this kind of differential allocation of feeding effort may not be a means of increasing individual fitness by favouring those chicks that are themselves presumed to be fit, it may simply be a means by which a pair of birds can make the process of raising a family more efficient.

In the case of redstarts, and specifically the Black Redstart *Phoenicurus ochruros*, Tudor Draganoiu and his colleagues have shown that the members of a breeding pair each preferentially feed some members of the brood while paying little if any attention to the others. Often they observed males to feed fewer of the chicks than females, and in some cases the division of labour was such that the female parents fed three times as many chicks as the males. Although the reason for this behaviour has not been determined in this species, Draganoiu and his team have been able to show that birds are able to recognize the chicks that they will feed. By observing the response of adults to recordings of chick begging calls they have shown that adults are able to discriminate between the begging calls of individual chicks.

> **Key reference**
>
> Draganoiu, T.I., Nagle, L., Musseau, R., and Kreutzer, M. (2006) In a songbird, the Black Redstart, parents use acoustic cues to discriminate between their different fledgelings. *Animal Behaviour* **71**, 1039–1046.

Box 5.7   Helpers at the nest

Raising a family can be a chore, and evidence that two parents are better able than one to do the job is commonly cited in the scientific literature. However, there are cases when even two parents are not sufficient, and in these situations it is common for bands of birds to work together to raise a brood cooperatively. In some cases, a single breeding pair is assisted in rearing a brood by independent young from prior breeding attempts. In others, a number of pairs breed together in a colony system, sharing food-finding and defence roles. From the point of view of understanding some of the factors driving the development of helping as a strategy, the work of Jan Komdeur and colleagues on the Seychelles Warbler *Acrocephalus sechellensis* is worthy of special mention. In the mid 1960s the world population of this species was as low as 26 pairs and was confined to just one of the islands in the Seychelles archipelago—Cousin Island. Effective habitat management by the International Council for Bird Preservation (ICBP) increased the population to around 300 birds by the early 1980s, occupying around 120 territories (this

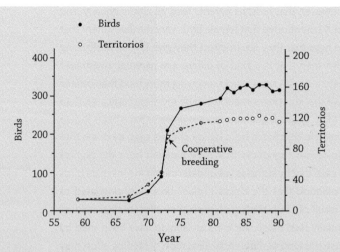

Figure 5.10 The growth of the population of Cousin Island Seychelles Warblers. From Komdeur, J. (1992) Importance of habitat saturation and territory quality for evolution of cooperative breeding in the Seychelles Warbler. *Nature* **358**, 493–495.

is thought to be the carrying capacity of the island, i.e. the maximum number of territories available). Figure 5.10 illustrates the growth of the population and highlights one of the most interesting points about it—there are many more breeding age birds than there are territories, and at about the time that the island became saturated some birds stopped dispersing when they became independent and chose instead to remain on their natal territory as helpers.

These helpers assisted their parents in incubation (females only), chick feeding, territory defence, and predator mobbing. They were most often found on higher quality territories; the ones that were particularly rich in insect food and so could support a larger population of birds. As territories became available, helpers had a choice—stay on at home or move out to occupy the vacant space. In many ways they acted in a manner similar to that predicted by the PTM. When the first vacancy to arise was in a poor quality territory they stayed home and waited until something better 'reached the market'. However, if a high quality territory became available first, they moved onto it. Essentially this seems to be because considered over a reproductive lifetime the immediate term benefit of taking up a low quality opportunity and then staying on that territory for the remainder of one's life is less

than the benefits to be gained by waiting until the high quality territories come along. In fact in some cases helpers on low quality territories have actually been shown to reduce the productivity of their parents, presumably by competing with the brood for food.

Remember all of the time that a helper is assisting in the raising of its siblings it is in effect helping to propagate genes that it shares with them. In this way, a young bird that delays dispersal and breeding is still making an indirect contribution to its own genetic fitness. Recently Komdeur and his coworkers have added a new twist to this tale. It turns out that grandparents are helpers too! Over a 24 year observation period almost 14% of breeding females were deposed by one of their younger relatives. Rather than disperse and become non-breeding floaters (birds without a territory) 68% of these grandmothers stayed on and assisted their offspring in the raising of the next generation, thereby continuing to gain an indirect fitness benefit themselves.

Confirmation of the generality of this sequence of helper behaviour development has come from recent conservation efforts that have introduced small numbers of the warblers to adjacent Aride (1988) and Denis Islands (2005). In the case of Aride no helping was recorded until the island had reached its carrying capacity.

References

Brown, J. (1987) *Helping and Communal Breeding in Birds; Ecology and Evolution*. Princeton University Press, Princeton.

Richardson, D.S., Burke, T., and Komdeur, J. (2007) Grandparent helpers: the adaptive significance of older, postdominat helpers in the Seychelles Warbler. *Evolution* **61**, 2790–2800.

Imprinting and independence

During the nestling and postfledging stage, before they become fully independent, young birds imprint upon their parents. This very specific learning process serves to 'fix' the species identity of the individual and set it on a behavioural path that will persist into adulthood. As we saw in the case of the sensitive period of song acquisition, the imprinting period is a short and discrete period in the young bird's life. As an example of the significance of imprinting, consider the mating preferences of adult birds that were imprinted upon a 'parent' of a species other than their own. If eggs of the Zebra Finch *Taeniopygia guttata* are cross-fostered to be hatched and reared by the domesticated Bengalese Finch *Lonchura striata* parents, the resulting male Zebra Finches will preferentially court female Bengalese Finches and ignore females of their own species. For this reason birds that are hand reared as part of a conservation programme must be maintained in a carefully controlled environment and provided with suitable specific stimuli to ensure normal development. However, there is occasionally an advantage to inappropriate imprinting, and male raptors that have been hand reared specifically to imprint them upon a human trainer can be encouraged to mate and ejaculate with their trainer's gloved hand to facilitate the collection of semen to be used in artificial insemination programmes.

Eventually, if a chick has survived the competition with its siblings, the vulnerable period in the nest, and the risky business of fledging into a hostile environment, there comes a time when it must become independent. In some cases, birds remain in extended family groups (see Box 5.7), but in most cases they drift away from, or are chased away by, their parents to disperse and begin their adult life.

Summary

As a result of anisogamy the priorities of males and females differ—although of course both are primarily driven to pass on their genes. Consequently a wide range of mating/breeding strategies have evolved. Females choose males on the basis of song, display, resource provision, or genetic quality, and males compete for access to mates. Chicks manipulate their parents, but mothers and fathers may also manipulate the care that they provide.

Questions for discussion

1. Explain a range of mating strategies in terms of the motivations/needs of both males and females.
2. Why do birds sing?
3. How are offspring–parent relationships influenced by available resource levels and by the social activities of parent birds?

CHAPTER 6

Foraging and avoiding predators

A curious bird is the Pelican, its beak can hold more than its belly can!

D.L. Merritt, Nashville Banner, 1913.

Birds need to eat and drink. The basic principles of foraging for food and water are common to all species be they a predatory owl, a herbivorous goose, a generalist omnivore like the crows, or a highly specialized feeder like the nectivorous humming-birds. Food or water has to be found, captured, or obtained, and then processed and ingested. At the same time, a foraging individual has to make decisions about food quality, perhaps choosing one item over another, make decisions about how much to eat and in some cases how much to store, and make decisions about when to share with flock mates and when to defend the resource. Think about this next time you watch a feeding bird—their behaviour may not be as simple as it appears!

Chapter overview

6.1 Foraging
6.2 Optimal foraging
6.3 Risk and foraging
6.4 Predator avoidance

6.1 Foraging

Finding food and capturing prey

In Chapter 3 we saw that birds such as chickadees and titmice are able to re-locate previously stored food by impressive acts of memory. In many cases, however, foraging involves the location and acquisition of previously unidentified food sources and involves the use of the full range of senses.

Flight path: foraging can involve navigation and spatial memory. Chapter 3.

Key reference

Montgomerie, R. and Weatherhead, P.J. (1997) How robins find worms. *Animal Behaviour* **54**, 143–151.

Birds have highly developed eyes and high visual acuity, so it should come as no surprise that many of them rely upon sight to locate prey. Some, such as the hawks and owls, have forward-pointing eyes which facilitate binocular vision—essential for a bird that grabs moving prey. Others have eyes more towards the sides of their heads and have been forced to compromise between binocular vision for prey capture and monocular vision, with a wider field of view, for predator avoidance. These birds often have to cock their head from side to side to get an accurate fix on their target prior to pecking. Of course there are occasions when sight is not sufficient. Nocturnal owls can see by moonlight, but they rely to a greater extent upon their hearing to locate prey. Similarly it has been demonstrated experimentally by Robert Montgomerie and Patrick Weatherhead that although foraging American Robin *Turdus migratorius* do use visual cues when hunting for worms in leaf litter and soil, they are less able to locate their prey when it is immobile or if the sounds made by crawling worms are masked by white noise. This suggests that robins use auditory cues to find worms.

Wading birds feeding on soft sediments are able to see prey and visual cues to the presence of prey (worm casts and burrow mouths) during the day or under a bright moon. However, the pressures of feeding between the tides mean that these birds often have to feed when they are unable to see well, and of course long-billed birds such as curlews cannot see the prey that they seek when they probe deep into soft mud. In such cases waders feel for their food. The tip of the wader beak is sensitive to touch and able to discriminate prey from non-prey. Smaller billed waders, sandpipers and plovers, use a similar tactic when searching for surface prey in the dark. By day they can see their prey and dart across the sand to snatch it up with a single well-aimed peck. At night they hunt by 'stitching', repeated rapid jabs at the sand/mud as they walk across it (like the stabbing of a sewing machine needle—hence the term stitching), occasionally swallowing food that they encounter.

Nocturnal kiwi feed largely on burrowing earthworms which they capture by probing into soft earth with their long beak in much the same way as the long-billed waders. In this case, however, smell rather than touch seems to be the important sense. Whereas the nostrils of most birds are situated at the base of the beak, those of the kiwis are close to the tip. It seems that they find their prey by sniffing out burrows. Similarly, it has recently been shown that some of the oceanic albatrosses and petrels are stimulated to follow the scent of pyrazine. This is a chemical compound that is released when plankton and krill are eaten by fish or other birds. This might explain how some species are able to home in on patchily distributed but locally superabundant swarms of krill.

Key reference

Nevitt, G., Reid, K., and Trathan, P. (2004) Testing olfactory foraging strategies in an Antarctic seabird assemblage. *Journal of Experimental Biology* **207**, 3537–3544.

Sharing information

Although petrels may follow pyrazine slicks to locate food, if the krill are under attack by other surface-feeding birds the visual stimulus of a feeding frenzy is likely to be a very strong cue to any bird in the area as to the whereabouts of that food patch. The idea that birds might use the sight of the success of other foraging birds as a cue to the hunt is encapsulated in the 'information transfer hypothesis' which simply predicts that animals should act upon information from others for their own benefit. In the case of a petrel flying towards a mob of feeding birds, there is of course no suggestion that the cue is an intentional signal designed to recruit others, but in some cases intentional communication does seem to be used to recruit flock mates to good feeding areas.

During observations of a colony of Osprey *Pandion haliaetus*, Erik Greene noted that whilst birds returning to the colony after an unsuccessful fishing trip flew directly to their nest or favoured perch (Plate 24), birds that had been successful (they were carrying a fish) did not. Instead successful birds performed an elaborate undulating display flight accompanied by a persistent call. Greene hypothesized that if these birds were intentionally advertising their success then their colony mates should take advantage of the information provided to them and fly out to hunt more often when they are 'told' that fish are available, and fly most often in the direction from which the successful hunter approached the colony. This is exactly what he recorded happening.

There is an added sophistication to this behaviour. From Figure 6.1 it is clear that when the returning hunter carries an alewife, pollock, or smelt into the colony there is an increase in the number of birds setting out on a hunting trip (compared with the number setting out with no information or with negative information). The figure also shows that when the hunter returned with a winter flounder, colony mates did not respond. This is because unlike the shoaling alewife, pollock, and smelt, winter flounder is a solitary fish.

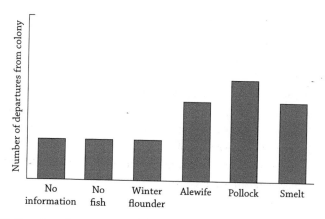

Figure 6.1 Departures from an osprey colony are more common when birds have observed a successful colony mate returning with a shoaling fish. From Greene, E. (1987) Individuals in an osprey colony discriminate between high and low quality information. *Nature* **329**, 239–241.

Foraging flocks

Sharing information and working together can enable birds to forage more efficiently as a flock and to have a higher level of success than they would were they to forage alone. For example, experiments with captive flocks of fishing Black-headed Gull *Larus ridibundus* have shown that as flock size increases the likelihood of an individual catching a fish also increases (Figure 6.2). This is probably because fish fleeing from lots of predators are likely, in their confusion, to blunder into the beak of one of them rather than because the birds are actively working together.

In the case of the Harris Hawk *Parabuteo unicinctus*, however, real cooperation is the order of the day. Groups of hawks gather in the early morning and then spend the day actively searching for prey in an extended flock. Individual hawks leap-frog over one another as they search for rabbits and other small mammals (see Figure 6.3).

Once one hawk has found a prey animal, the others in the group converge upon it and then they work together to move it into a position from which one or more of them can launch an attack. Sometimes they drive the animal into the open and then several birds pounce upon it from different directions. If the prey takes to cover, one bird will follow it to flush it out towards the others who are waiting in ambush. When a kill is made, all members of the group share in the meal and it would appear that a single rabbit is sufficient to satiate several hawks. Solitary Harris Hawks are rarely seen hunting, and so cooperative hunting would appear to be the norm in this species. Indeed Bednarz suggested that on the basis of his

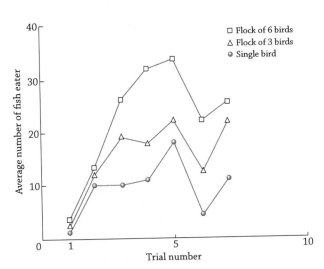

Figure 6.2 Birds in larger flocks are more successful when hunting than are solitary birds. From Göttmark, F., Winkler, D., and Andersson, M. (1986) Flock-feeding on fish schools increases success in gulls. *Nature* **319**, 589–591.

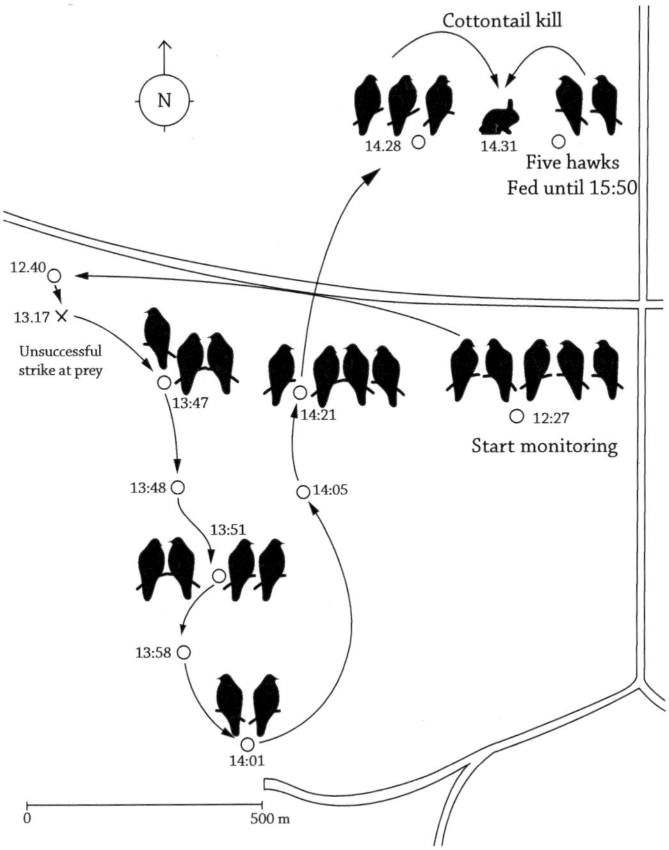

Figure 6.3 Sequence of movements of Harris Hawks during an ultimately successful hunt. Perched hawks indicate the number of birds recorded in the group at that time and location. From Bednarz, J.C. (1988) Cooperative hunting in Harris' Hawks (*Parabuteo unicinctus*). *Science* **239**, 1525–1527.

observations, it is probably the only way that Harris Hawks can survive in the harsh desert environment of southeastern New Mexico.

Do herbivores cooperate?

It is easy to see how by confusing prey or cooperating to overcome it gulls and hawks are coordinating their behaviour to maximize their individual success. But can the same ever be said of herbivores? Plants do not need to be overcome, nor do they try to escape (although of course it would be wrong to think that they were defenceless—just grasp a nettle!). Later in this chapter we will see that flock formation in herbivorous birds may be a predator avoidance strategy, but here I would like to consider the possibility that it is also a means by which birds can maximize their individual success in terms of the quality of the food that they have available to them.

Key reference

Prins H.H.Th, Ydenberg, R.C., and Drent, R.H. (1980) The interaction of Brent Geese *Branta bernicla* and Sea Plantain *Plantago maritime* during spring staging: field observations and experiments. *Acta Botanica Neerlandica* **29**, 585–596.

In Europe, Brent Geese *Branta bernicla* spend the winter on temperate estuaries and salt marshes where they feed on coastal vegetation, and in particular on Sea Plantain *Platago maritima*. The plants that they eat are not particularly nutritious but they are abundant, and supplies are replenished through re-growth within a few days of grazing. New shoots are more nutritious than old growth. So we would predict that the best strategy for a bird is to feed on a patch and then to stay away from it until re-growth has occurred—but not to stay away so long that the growth becomes old and tough. Of course this can only work if all of the members of a flock coordinate the way that they re-graze particular patches. If they do not, then birds would probably always attempt to return a little too early to beat their competitors. Observations of grazing geese made by Herbert Prins and his coworkers have shown that on Dutch salt marshes Brent Geese do cooperate to re-crop areas on a 4-day rotation. They have also shown (by experimentally clipping the plants) that 4 days is the optimal time to return in order to maximize the amount of new growth material available to them. So it would seem that cooperation is as much of an advantage to herbivores as it is to predators.

6.2   Optimal foraging

The geese that we have just considered are behaving as 'intelligent foragers' in an optimal fashion. That is to say their behaviour is exactly what we would predict if they are attempting to maximize the benefits (food intake) of their activity at the same time as minimizing the costs.

Many species of coastal bird feed upon intertidal molluscs. Some like the Eurasian Oystercatcher *Haematopus ostralegus* use their beaks to open their prey, and as an aside they have evolved two distinct morphologies and behaviours to do this. Some oystercatchers have sharp pointed beaks that they use to stab between the valves (shells) of their prey and then prise or twist the shells apart to access the flesh. Others have a heavy blunt-ended beak that they use as a hammer or chisel to smash open the shells of their prey. Gulls and crows, however, use a different strategy. Having selected a potential prey item, they carry it into the air and drop it to shatter on the rocks below.

During a study of the whelk-dropping behaviour of Northwestern Crows *Corvus caurinus*, Reto Zach noted that birds were very selective when choosing a whelk to drop. Although whelks ranging from around 1.5 to 5 cm were available in the environment, the crows preferentially selected larger whelks for dropping (see Figure 6.4A). He also noted that they consistently dropped whelks from a height of around 5 m and tended to have preferred drop zones (which were littered with broken shells).

Through a series of experiments, Zach ruled out the possibility that smaller whelks were unpalatable (crows were equally likely to consume the flesh of smaller and larger

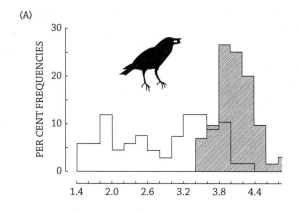

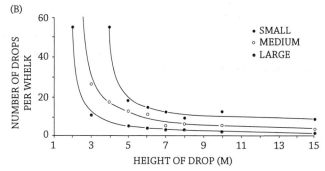

Figure 6.4 (A) Although whelks ranging in size from 1.4 to 5 cm in length were available (white bars) crows preferentially selected rarer, larger whelks (black bars). From Zach, R. (1978) Selection and dropping of whelks by Northwestern Crows. *Behaviour* **67**, 134–148. (B) Large whelks require fewer drops to break their shells than do smaller whelks. From Zach, R. (1979) Shell dropping: decision-making and optimal foraging in Northwestern Crows. *Behaviour* **68**, 106–117.

whelks when presented with a choice). He also found that larger whelks broke more easily than smaller ones when dropped (Figure 6.4B) and unsurprisingly provided the greatest reward (having most flesh). He also found that shells were more likely to break if he dropped them in the crows' preferred drop zones than if he dropped them elsewhere. But most significantly he calculated that by making multiple drops from 5 m (rather than fewer from a higher height) the birds maximized their net energy return when they ate the whelk (net energy return remember is the amount of energy obtained from the food minus that spent finding it and making the flights to break it). He showed that the birds were foraging in an optimal fashion.

Feeding territories

As we saw in Chapter 5 one of the functions of a territory is to provide the resources a bird or a pair or group of birds need to raise their young. But many birds retain their territories (or acquire new ones) outside of the breeding season. In some cases these non-breeding territories occupy all or most of the same area as the breeding territory, and it is likely that they are retained between years to facilitate breeding. Others are only held outside of the breeding season and so their prime function seems to be to provide the holder with access to sufficient food. In this respect some territories may be quite small and persist for a very short period.

For example, Sanderling *Calidris alba* congregate on beaches in feeding flocks, and large numbers of birds can often be seen feeding alongside one another. However, these diminutive waders may, under certain conditions, defend small feeding territories. Myers and coworkers made observations of the flocking and territorial behaviour of sanderling feeding on the isopod *Exirolana linguifrons* on a sandy beach. Their results (Figure 6.5) indicate that the decision to defend a territory depends upon the amount of food available. They found that when prey were scarce birds fed alongside one another; presumably it would be impossible for an individual to defend an area large enough to provide sufficient food. Similarly, when food is superabundant there is presumably no need to defend the resource and birds do not have territories. However, at intermediate prey densities (and the thresholds involved are quite narrow) the birds do become territorial. Here the costs of defence are outweighed by the benefits of controlling access to the available food.

The actual energetics of the trade-off involved in this kind of territorial defence have been investigated in the case of the old world sunbirds and new world hummingbirds. Both rely to a great extent upon energy-rich nectar, and nectar availability is a key determinant of territory size. For example, in a now classic study of the foraging energetics of Golden-winged Sunbird *Nectarinia reichenowi*, Frank

Key references

Gill, F.B. and Wolf, L.L. (1975) Economics of feeding territoriality in the Golden-winged Sunbird. *Ecology* **56**, 333–345.
Geerts, S. and Pauw, A. (2009) African sunbirds hover to pollinate an invasive hummingbird-pollinated plant. *Oikos* **118**, 573–579.

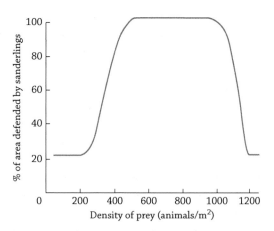

Figure 6.5 Sanderling defend feeding territories only at intermediate prey densities. From Myers, J.P., Connors, P.G., and Pitelka, F.A. (1979) Territory size in wintering sanderlings: the effects of prey abundance and intruder density. *Auk* **96**, 551–561.

Gill and Larry Wolf showed that in defended patches of flowers nectar levels were higher than in undefended patches. They calculated that birds defending patches expended around 26 kJ of energy during an 8 hour observation period, whereas those birds without a territory spent around 32 kJ.

Sunbirds generally feed on nectar whilst perched on flowers, but recently the invasion of parts of South Africa by a new world plant, *Nicotina glauca*, which is usually pollinated by hovering hummingbirds, has resulted in a startling behavioural development. Sjirk Geets and Anton Pauw have reported that the members of a community of Double-collared Sunbirds *Cinnyris afer* and Malachite Sunbirds *Nectarinia famosa* have started to hover to obtain nectar from *Nicotina glauca* (and to pollinate the flowers in the process) (Plate 25). The effect of this has been that areas with *Nicotina glauca* support larger numbers of sunbirds than do areas with only native flower species present, and sunbirds delay their seasonal migration, spending longer in *Nicotina*-rich areas than would previously have been the case.

> **Key reference**
> Geerts, S. and Pauw, A. (2009) African sunbirds hover to pollinate an invasive hummingbird-pollinated plant. *Oikos* **118**, 573–579.

6.3 Risk and foraging

So far we have considered foraging as an isolated behaviour, but of course in reality finding food is just one of a number of concurrent activities undertaken by an individual. The individual doesn't just have to decide what to eat, it also has to decide when to eat it. This might involve storing food during a glut as a way to survive a future shortage, like the Marsh Tits discussed in Chapter 3. It might also involve weighing up the risks of feeding against the risks of not doing so (which might ultimately be starvation). For example, many birds are reluctant to forage far from cover. Titmice, sparrows, and juncos have all been shown to prefer to feed closer to cover into which they might flee when disturbed or attacked, and evidence exists that birds are able to modify their behaviour in response to changing levels of risk. For example, Jukka Suhonen found that during periods of relatively low predation risk, Crested Tits *Parus cristatus* foraging in trees utilize both high and low branches and feed close to the trunk and outwards to branch tips. However, during years when small mammal populations were particularly low and Pygmy Owls *Glaucidium passerinum* changed their behaviour to hunt birds, the tits responded by concentrating their own foraging closer to tree trunks where they would be less likely to be attacked.

On the other hand, there are situations in which birds will take risks, or at least appear to do so. For example, Stephen Lima and his colleagues have shown that towhees and song sparrows feed further from cover than would be expected if they were behaving in a way that would minimize the time taken to flee to safety. This is possibly because cover is paradoxically also a risk—many predators make ambush attacks from cover and in such cases feeding in the open might be an advantage.

> **Flight path:** mixed species flocks divide up a resource to minimize competition. Chapter 7.

> **Key reference**
> Suhonen, J. (1993) Predation risk influences the use of foraging sites by tits. *Ecology* **74**, 1197–1203.

> **Key references**
> Lima, S.L., Wiebe, K.L., and Dill, L.M. (1987) Protective cover and the use of space by finches: is closer better? *Oikos* **50**, 225–230.
> Lima, S.L. (1988) Initiation and termination of daily feeding in dark-eyed juncos: influences of predation risk and energy reserves. *Oikos* **53**, 3–11.

Lima's observations suggest that the finches he studied optimize their behaviour by arriving at a compromise between the costs and benefits of proximity to cover.

It has also been shown that when resources are scarce, or competition close to cover is high, subordinate birds (presumed to be those least able to compete) will take increased risks and feed in the open. In another study, Lima has shown that those Dark-eyed Juncos *Junco hyemalis* starting their day with the lowest energy reserves (stored fat) are more likely to start feeding in low light when predators are more difficult to locate and both nocturnal and diurnal predators may be active. Birds with sufficient fat stores are more likely to wait until light levels and, therefore relative safety from predators, increase.

6.4   Predator avoidance

It is likely that some level of risk is inevitable and that at some stage of their life cycle all birds face the threat of predation. In the preceding section of this chapter, however, we saw that by adjusting their behaviour, birds are able to adjust the level of risk to which they are exposed. In this section I would like to consider predator avoidance in more detail.

Camouflage

Many birds are difficult to see. Some simply skulk and remain inconspicuous; others have evolved plumage that is camouflaged. My personal experience of camouflage is one of frustration and amazement. I know first hand how difficult it can be to find an inconspicuous Leaf-warbler amongst a canopy of leaves or a Straw-coloured Bittern on the fringes of a reed bed, and on countless occasions I have almost trodden on a Ringed Plover *Charadrius hiaticula* chick as it crouches pebble like on a rocky beach (Plate 26). But what evidence is there that camouflage benefits birds by protecting them from predators rather than from the gaze of a curious birder?

Esa Hutta and coworkers have approached this question from an unusual angle. Rather than attempting to show that camouflage reduces predation risk, they have demonstrated that the inverse is true, i.e. that brightness increases predation risk. To assess vulnerability to predation they looked at the relative proportions of the bright- and dull-plumaged passerine remains found in and around Sparrowhawk *Accipiter nisus* nests and known plucking posts over a 30-year period. Their analysis revealed that as they predicted, brightly plumaged species were over-represented in the hawks' diet and were presumably therefore easier for hunting hawks to catch than were dull-plumaged, camouflaged, species.

In another study involving Sparrowhawks and their prey, Götmark and Olsson have dramatically demonstrated the cost of increased conspicuousness. To do this

Concept
Camouflage

When we find it difficult to see a bird because its coloration blends with that of its environment we think of it as being camouflaged. However, it is important to remember that what one animal sees as being camouflaged another may not. Birds are able to see reflected UV light, and species that look dull to us may therefore be very bright to other birds.

Key reference

Hutta, E., Rytkönen, S and Solonen, T. (2003) Plumage brightness of prey increases predation risk: an among-species comparison. *Ecology* **84**, 1793–1799.

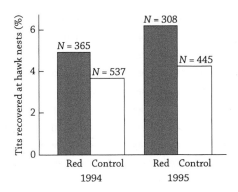

Figure 6.6 The percentages of conspicuous red-painted and inconspicuous yellow-painted Great Tits recovered from the nests of Sparrowhawks during two seasons. *n* indicates the total number of each class of tits painted in each season. From Götmark, F. and Olsson, J. (1997) Artificial colour mutation: do red-painted tits experience increased or decreased predation? *Animal Behaviour* **53**, 83–91.

they manipulated the plumage of young Great Tits while they were still in the nest. Great Tits have striking yellow, black, white, and greenish plumage which might sound quite conspicuous, but is in fact relatively inconspicuous against the mottled light and shade of a woodland canopy. Some of the birds were made less conspicuous by having the white feathers on their cheeks, wings, and tails painted yellow. Others were made more conspicuous by having their white feathers painted red. All of the chicks were fitted with numbered metal rings to facilitate individual recognition and then allowed to fledge normally.

The main predators of fledgling Great Tits are Sparrowhawks which time the hatching of their own eggs to coincide with the annual titmouse glut. Two weeks after the tits fledged, the researchers used a metal detector to search for their rings in the area of Sparrowhawk nests and plucking posts. The results of their searches (Figure 6.6) revealed that red-painted birds were 38% more likely to be predated then yellow-painted birds, evidence that conspicuousness increases predation risk and that inconspicuousness (camouflage) decreases it.

Predator distraction displays

Plovers and other ground-nesting birds are generally well camouflaged, as are their eggs and mobile young (see Plate 26). However in terms of avoiding predation the plovers are better known for their unusually conspicuous behaviour and habit of drawing attention to themselves, behaviour which might at first seem paradoxical in the face of a predation threat.

Plover nests are typically on open ground, and a vigilant bird will detect a threat (a fox or mustelid typically) when it is still some distance away. Having identified a predator, the plover discretely leaves its nest and walks a little distance from it. The bird will then employ one or more of a number of antipredator strategies. Plovers that have been incubating eggs often sit and pretend to incubate, allowing the predator to find and flush them. The bird then flies to safety leaving the hapless predator to search in vain for non-existent eggs.

In other cases if the plover is in long grass and cannot be easily seen by the hunter it will scurry away uttering a high-pitched squeaking sound effectively impersonating a small rodent. Typically the predator will give chase and when it has moved a sufficient distance from the nest or young the plover will fly to safety. If the predator seems not to want to follow it the plover will often run towards it calling loudly and then at the last moment darting away, usually followed by the predator which is by now well and truly hooked.

There is, however, another behaviour for which the plovers are renowned, and that I can confirm is highly effective having fallen for it myself on numerous occasions. This is the broken-wing display. Having identified a threat, the plover will stand in a conspicuous position and draw attention to itself with a repetitive pipping call. At the same time it will lower one wing (feigning an injury) and start to slowly 'limp' away. Presented with such an easy target the predator gives chase and the plover, following an irregular path, leads it away from its nest or chicks (chicks crouch immobile in response to an alarm call from their parent). If the predator appears to lose interest the plover exaggerates its plight even more, often falling to the ground and flailing an apparently useless wing (see Plate 27). But as soon as the predator comes close to grabbing its prey or has moved a sufficient distance from the nest, the plover utters an almost mocking call and flies to safety.

Tonic immobility

If these strategies fail and chicks are found, they too have a ruse by which to affect their escape. When they are picked up by a predator the chicks of ground-nesting birds exhibit what is referred to as tonic immobility—basically they become limp and play dead. Because such chicks tend to come in groups, predators often drop 'dead' chicks in the hope that they will find another. Of course when they subsequently return to re-claim their prize they find it gone. Tonic immobility can be induced if a chick is presented with a stimulus mimicking two forward-pointing eyes (like those of a predator) and can be sustained for up to 30 minutes if the stimulus is not removed. During this time the chicks will regularly open one eye a little just to check whether the danger is still present.

Alarm calls

When they detect a predator, individuals of most, if not all, bird species produce an alarm call. One of the functions of such calls, and indeed of other forms of alarm behaviour, is probably to inform the predator that it has been detected and that having lost the element of surprise an attack is unlikely to succeed. In this way alarm calls could act as a predator deterrent. The other main function of alarm calls is to provide flock mates with the vital information that a predator has been detected and to stimulate them to take appropriate antipredator actions. Such responses might include becoming silent and inconspicuous (particularly in the case of nestlings and dependent young birds), fleeing, or somewhat paradoxically making one's presence known to the predator and even attacking it, as we shall see later in this chapter when we consider mobbing behaviour.

Walking in English woodland, one of the most common calls that I hear is a high-pitched *seet*. Usually this is the alarm call of a European Robin *Erithacus rubecula* that has been disturbed by my presence. Through this call the robin shares with the wider community of woodland birds the information that I am there and that I am a potential threat (of course I'm not really!). This multispecies sharing of information is enhanced because the alarm calls of most if not all of the members of the community have evolved to become very similar indeed (see Figure 6.7A). Typically they are relatively drawn-out calls with a rather narrow frequency range, making them very difficult for a predator to locate accurately. This of course is an advantage to the birds in that a predator cannot attack a target that it is unable to locate.

On the one hand, the alarm call system would therefore appear to be a relatively simple one. However, recent research has revealed that in fact it is often used in quite a sophisticated way. For example, it has been shown by Christopher Templeton and his coworkers that Black-capped Chickadees *Poecile atricapilla* vary their alarm call in response to the identity of the predator that they have detected. The chickadees are named for their *chick-a-dee-dee* call which is used as an alarm/mobbing call when predators are detected. By exposing chickadees to models of a range of potential predators the researchers found that they varied their call, primarily by adding in more repeats of the terminal *dee* note, in response to increasing predator size (see Figure 6.8A). Furthermore, by playing back a range of chickadee alarm calls, they also showed that conspecifics vary their responses to different alarm calls, taking note of the content of the call and decoding the potential risk that the predator poses. So, for example, they respond most strongly to smaller predators which are themselves more likely to attack chickadees than are larger predators (Figure 6.8B).

Templeton and Greene have also shown that members of other species in the forest community which includes the Black-Capped Chickadee are able to decode the messages transmitted between alarm-calling chickadees. For example, Red-breasted Nuthatches *Sitta canadensis* respond to the alarm calls of neighbouring Black-capped Chickadees and take note of more than the presence of a threat. In response to chickadee alarm calls stimulated by the presence of a Northern Pygmy

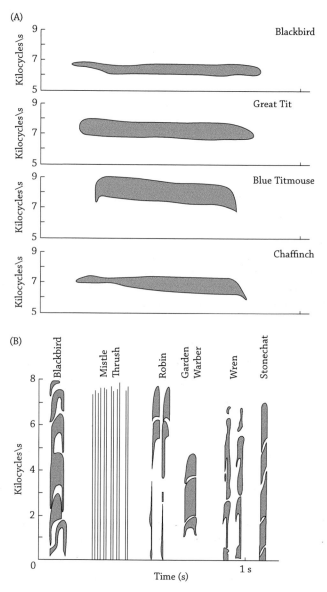

Figure 6.7 Sonograms of alarm calls (A) and mobbing calls (B) of a range of coexisting European woodland birds. Note that whereas alarm calls typically have a narrow frequency range, making them difficult to pin-point, mobbing calls have a wide frequency range and are easily locked onto by predators. Note also the convergence in the evolution of both types of call. From Marler, P. (1959) Developments in the study of animal communication. In: Bell, P.R. (ed.) *Darwin's Biological Work*. Cambridge University Press, Cambridge, pp. 150–202.

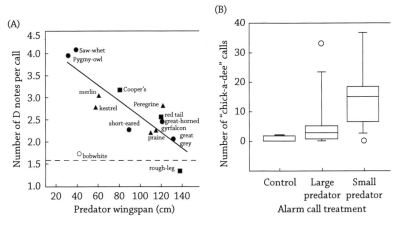

Figure 6.8 (A) The intensity of alarm call (number of *dee* (D) notes in the call) clearly increases as predator size decreases, and chickadees clearly respond most strongly to hearing alarm calls that are a response to the greater threat posed by the presence of a smaller predator (B). Adapted from Templeton, C.N., Greene, E., and Davis, K. (2005) Allometry of alarm calls: Black-Capped Chickadees encode information about predator size. *Science* **308**, 1935–1937.

Owl *Glaucidium gnoma*, a species which also predates nuthatches, they responded with high intensity alarming and mobbing. But when the chickadees' alarm calls were a response to the presence of a Great Horned Owl *Bubo virginianus*, a species which is not known to hunt nuthatches, their antipredator response was far less strong.

There is, however, evidence that heeding the warnings of others may sometimes be a costly behaviour. Fork-tailed Drongo *Dicrurus adsimilis* are insectivorous and tend to take prey by hawking it from the air. They will, however, also take prey from the ground, particularly when it has been disturbed by a ground-foraging bird, a form of kleptoparasitism. When foraging alone, drongos will alarm call in response to the threat posed by an aerial predator, but they rarely call in response to a terrestrial one. However, when in the company of ground-foraging Pied Bablers *Turdoides bicolor* the drongos take on a sentinel role and will alarm in response to both aerial and terrestrial predators. This of course benefits the babblers who, relying upon their drongo look-outs, reduce the level of their own vigilance (spending more time foraging) and respond to the drongo alarm by fleeing to the safety of deep cover. The intriguing question of course is why do the drongos go to this trouble? Amanda Ridley and her colleagues have shown that in some cases they use this relationship to their own advantage. Occasionally they will alarm dishonestly (i.e. when there is no predator present), causing the fleeing babblers to drop or leave their own prey which is quickly snapped up by the hungry drongo. This dishonesty can persist because it is a strategy that the drongo uses sparingly and the babbler is unable to call its bluff.

Key reference

Templeton, C.N. and Greene, E. (2007) Nuthatches eavesdrop on variations in heterospecific chickadee mobbing alarm calls. *Proceedings of the National Academy of Sciences, USA* **104**, 5479–5482.

Key reference

Ridley, A.R., Child, M.F., and Bell, M.B.V. (2007) Interspecific audience effects on the alarm-calling behaviour of a kleptoparasitic bird. *Biology Letters* **3**, 589–591.

**Concept
Sentinels**

Single and mixed-species flocks often include individuals who take on the role of sentinel. These individuals act as a look-out for the flock, devoting more time to vigilance and therefore less time to foraging than their flock mates.

Mobbing

Whilst alarm calls appear to have evolved to have a frequency range which reduces the chances that a predator will be able to locate the caller, mobbing calls, in contrast, have evolved to be easily located (Figure 6.7B). As a strategy, mobbing depends upon the predator knowing that it has been detected and that an attempt to hunt will probably result in failure. Mobbing is most commonly observed when passerine birds respond to the presence of an aerial predator. It may begin with a single mobbing bird, calling and diving at the predator repeatedly, perhaps even striking it. As the behaviour continues, other birds are recruited and very quickly a small flock of birds will form, all harassing the unfortunate hunter. Typically, having lost the element of surprise or because it just cannot take any more, the predator will move on to hunt elsewhere. Numerous studies have demonstrated that mobbing is a successful strategy, and most birders will have seen it at work. In fact learning to recognize mobbing behaviour can be a great way to detect predators that you might otherwise have overlooked!

Mobbing does, however, have a potential cost, and there are reports of predators taking a mobbing bird that came too close. This is perhaps why birds tend to cooperate and mob in flocks. Brown and Hoogland, in a comparative study involving the mobbing behaviour of solitary and colonial species of swallow, have shown that solitary mobbers are forced to take greater risks (come closer to the predator) than are mobbers in flocks (Figure 6.9).

Flocks and colonies

Lowering individual risk and increasing the chance of success in mobbing birds is just one of the antipredator advantages of group living and/or colonial breeding. Whilst being part of a large and conspicuous group might on the face of it seem

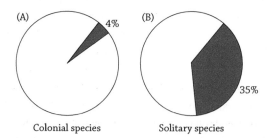

Figure 6.9 The percentage of mobbing actions that are high risk (shaded segments) is lower for colonial swallows mobbing as part of a flock (A) than it is for solitary birds (B). From Scott, G.W. (2005) *Essential Animal Behavior*. Blackwell Publishing, Oxford, adapted from Brown, C. and Hoogland, J.L. (1986) Risk in mobbing for solitary and colonial swallows. *Animal Behaviour* **34**, 1319–1323.

to make a bird more obvious to predators, there is in fact safety in numbers. By breeding in huge colonies and synchronizing the hatching of their young, seabirds are able to increase the effectiveness of their mobbing behaviour to deter predators, and simply to swamp them with food. There are so many eggs and chicks available that even the hungriest of predator populations will make relatively little impact upon the prey population as a whole.

At a very simple level, a phenomenon known as the dilution effect comes into play as group size increases. If a bird in a flock of one is attacked by a predator it has a 100% chance of being the target of the hunter. A bird in a flock of two has a 50% chance, and one in a flock of 100 has only a 1% chance (all things being equal) (see, for example, Figure 6.10). This effect is clearly at work in the large seabird colonies described previously and it probably also explains the flocking behaviour of female sea duck such as the Eider who bring together their chicks into often quite large crèches during their first vulnerable days at sea. During this period the chicks are easy prey for hungry gulls and there is little that either mother or chick can do to deter an attacking bird. Instead they rely upon the dilution effect and the ability of a number of females working together to spot danger and respond to it (by diving and encouraging the chicks to dive) sooner than a mother on her own would.

Further evidence that flocking is an effective antipredator strategy has been provided by Will Cresswell who has carried out an exhaustive study of the relationship between hunting Sparrowhawks *Accipiter nisus* and Peregrine Falcons *Falco peregrinus* and their Redshank *Tringa totanus* prey. Cresswell has demonstrated that the larger a redshank flock, the lower the probability that a given individual will be attacked by predators (Figure 6.10A). The redshank in the winter population that he observed faced two basic pressures each day—the need to avoid being eaten and the need to gain enough energy themselves. To an extent the two are in part difficult to reconcile because a feeding redshank often has its head down and so when actively foraging is far less able to see approaching predators. However, through group membership individual birds are able to increase the interval between bouts of vigilance behaviour and so maximize feeding time (Figure 6.10B). This is because in a sufficiently large group there will by chance always be some individuals on the look-out. There is an additional level of flexibility in this system in that all members of the flock increase their vigilance levels in response to a heightened perception of risk—immediately after a predator has been seen in the area for example.

One might expect redshank that have spotted a predator to use an alarm call, and some do. Others, however, do not. How can this be explained? Cresswell observed that redshank did call more often when escaping from an obvious threat (a raptor attack) than they did when the cause of their alarm was not apparent (flocks often simply spook themselves). Furthermore, he also noted more alarm calls by fleeing birds that had been feeding in a habitat that was visually obstructive (i.e. fellow flock members were not easy to see) than on an open mud flat. In the latter habitat they were much more likely to escape with a silent and direct fast flight to safety.

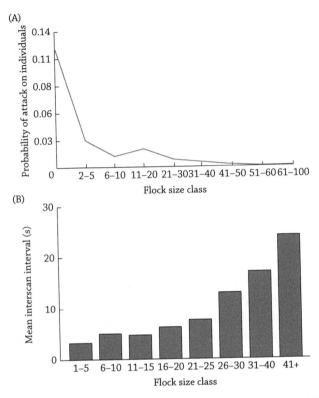

Figure 6.10 Increasing flock size results in a decreasing probability that any individual redshank will be the target of a predator (A), and allows individuals to forage for longer between bouts of vigilance (scans) (B). From Cresswell, W. (1994) Flocking is an effective antipredator strategy in redshanks, *Tringa totanus. Animal Behaviour* **47**, 433–442

Both escape behaviours had the same effect on the rest of the flock—causing all of the redshank in the area to take to the air. So, as was suggested previously in the chapter, one function of the alarm call does seem to be to coordinate an escape. Birds that do alarm call, potentially drawing attention to themselves, were no more likely than non-callers to be targeted by the raptor, so calling in itself is not a high risk strategy in this case. It is a highly effective strategy though because a coordinated escape presents a hunter with a mass of moving targets, making it impossible for it to pick out one to chase—this is often referred to as the confusion effect, another antipredator benefit of flocking.

Will Cresswell made a number of other important observations about the behavioural relationship between avian predators and their avian prey, and I would recommend that any interested reader take the time to read his numerous papers. But one of his observations I find particularly intriguing, and I will mention it briefly to round off this discussion. Remember that the redshank Cresswell observed were attacked by two different predators, Sparrowhawk and Peregrine Falcon, both of which use very different strategies to capture their prey. Sparrowhawk are a stealth hunter—typically breaking low from dense cover and hoping to snatch surprised prey from the ground. Peregrines on the other hand are pursuit hunters, stooping (a diving flight) towards prey from a height and at great speed and preferring to take air-borne prey. Cresswell noted that the redshank consistently responded to these predators in different ways. When attacked by a Sparrowhawk they bolt and escape with a low zig-zag flight pattern. In response to a Peregrine they freeze and crouch low to the ground. So in just fractions of a second birds are able to identify a potential threat, recognize the predator, evaluate its likely mode of attack, and take appropriate evasive action. An impressive feat, but an essential one—making a mistake would be fatal.

Summary

Foraging birds utilize a wide range of behavioural strategies when feeding and that behaviour continues to evolve to take advantage of new food resources. Essentially, however, they forage in an optimal fashion. The antipredator behaviour of birds is similarly diverse. Some species rely upon camouflage. Some cooperate to confuse a predator. Some take risks, exposing themselves to danger when mobbing hunters.

Questions for discussion

1. What are the advantages and disadvantages of life in a flock?
2. How do birds balance risk and hunger when feeding?
3. Is cooperation always advantageous?

CHAPTER 7

Populations, communities, and conservation

Man, however much he may like to pretend the contrary, is part of nature.

Rachel Carson, 1962

Although birds can be found in almost every corner of the globe, it is apparent to even the most casual observer that they are not distributed evenly; some places have more species of bird than others. Furthermore, whilst some individual species have very broad distributions, others are often found only in a particular type of place or even in just one very restricted area. To explain these patterns we have to consider some of the characteristics of bird populations and communities, particularly if essential conservation efforts are to be a success.

Chapter overview

7.1 Populations
7.2 Communities
7.3 Extinction and conservation

7.1 Populations

Bird populations vary in size dramatically. Populations of some species, such as the African Red-billed Quelea *Quelea quelea*, are counted in their millions, whilst populations of island endemics might include only a handful of individuals. Some populations appear to be stable (but fluctuate around a mean from year to year), others are growing, but many are a shrinking at such a rate that they are of immediate conservation concern. The factors governing both population size *per se* and trends in population size change are numerous and very often act in concert. In addition, the relationships between populations of species which form communities

Concept
Populations

can be defined as those members of a species which interact and have the potential to interbreed.

(see below) will mean that a shift in one population may have consequences for others and so have an impact at the community and ecological network level.

Life history strategies influence population growth

Some species have a naturally low capacity for population growth because of their particular life history strategy. Populations of larger birds with delayed maturation and small or infrequent clutches are typically slow to grow. For example, although considerable conservation efforts have been made on their behalf, the recovery of the Californian Condor *Gymnogyps californianus* has been painfully slow. These birds have a life span of some 50 years and take at least 6 years to reach sexual maturity. When they have found a mate and do begin to breed, their natural rate of productivity is very low. In 1987 only 22 of these magnificent birds remained, and all were captured and taken into captivity as the basis of a captive breeding programme. By 2009 birds had been returned to the wild and the conservation programme is a success, but the total population remains small, just 322 birds (including 172 in the wild).

Small passerines, on the other hand, have a much higher capacity for population growth. A pair of European Starlings, for example, can reproduce at 1 year old and can produce ten chicks per year (from two clutches of five eggs) for several years. As a dramatic example of the kind of population growth that can result from such a life history strategy, consider the fact that in 1890 Eugene Scheifflen introduced a founder population of between 60 and 110 (estimates vary) starlings into New York's Central Park and that in 2009 they have been estimated to be the most numerous bird in the USA (there are estimated to be more than 200 million of them). It took them around 50 years to colonize the USA from east to west, and as they did so they wreaked ecological havoc, outcompeting native species for access to nest cavities. Today in the USA starlings are considered a pest species, spreading zoonotic disease and causing significant agricultural losses. Where necessary, their numbers are controlled through culling by the application of a bird-specific pesticide (DRC1339) which is administered as poisoned bait. Paradoxically, in the UK the starling is a bird of conservation concern. Although not uncommon, the UK population has declined rapidly in recent years (Figure 7.1), possibly as a result of changes in agricultural practice that have reduced food availability and as a result of a loss of suitable nest sites as building regulations and standards reduce the number of cavities available in domestic roof spaces.

Population change

In spite of their intrinsic capacity for growth, some populations are in decline or have their growth limited in some way. Throughout the remainder of this chapter we will largely focus upon examples of such populations. We will consider a

Concept
Zoonoses

are diseases which can potentially be transmitted from animals to humans, several are known to be transmitted from birds to man directly, and others involve birds as an intermediate host or vector. See, for example Abulreesh, H., Goulder, R., and Scott, G.W. (2007) Wild birds and human pathogens in the context of ringing and migration. *Ringing and Migration* **23**, 193–200.

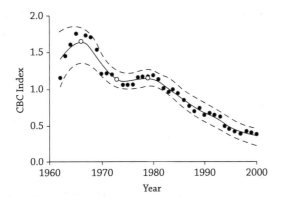

Figure 7.1 The population of Starlings in the UK declined dramatically over the period 1960-2000 (solid line and 95% CI—dotted lines). The decline in the index used (derived from the British Trust for Ornithology Common Bird Census) equates to a loss of more than 50% of the UK population. From Robinson, R.A., Siriwardena, G.M., and Crick, H.Q.P. (2005) Status and population trends of Starling *Sturnus vulgaris* in Great Britain. *Bird Study* **52**, 252-260.

range of factors which check population growth and then we will consider some of the ways in which population changes are involved in the structuring of bird communities.

Populations grow as a result of increased productivity and/or decreased mortality. This may be possible because resources become more abundant as a result of a climate shift, a human intervention, or even the misfortunes of another species, or it could be a result of a decrease in predator pressure, or in a conservation context because of an increase in protection. On the other hand, populations decline when productivity falls and/or mortality increases, perhaps because predator pressure increases, or a parasite or disease invades a population. It could be a result of increased competition as a result of a species introduction or a range expansion. Alternatively, it could be the result of shortage in resources because of poor weather or habitat loss. It could simply be that an extreme weather event kills lots of birds.

So, for example, it is not unusual for populations of small passerines to crash during a very cold winter. Similarly, populations of migrants are often dramatically reduced as a result of extreme climate events either during their migration or shortly after arrival at their breeding grounds. For example, in his excellent review of this phenomenon, Ian Newton reports that in 1993 a single tornado/storm event off the Louisiana coast resulted in the death of more than 40,000 migrating birds, and that unseasonal cold snaps in Europe have on occasion resulted in local hirundine population reductions of as much as 90%.

Populations that have suffered these crashes often rebound during the following years (assuming that mass mortality events are not repeated) because this mortality

Key reference

Newton, I. (2007) Weather-related mass mortality events in migrants. *Ibis* **149**, 453–467.

Key reference

Kluijver, H.N. (1966) Regulation of a bird population. *Ostrich* **6** (supplement), 389–396.

event may have lessened competition for nest sites/territories and food resources during the following spring, allowing a productivity increase for those birds fortunate enough to have survived.

Natural experiments like this have been duplicated under controlled conditions and similar effects measured. For example, Hans Kluijver has simulated an increase in breeding season mortality in Great Tit populations isolated on a Dutch island. He found that by removing a little over half of the breeding season population (adult birds and eggs/chicks) he was able to stimulate a significant increase in the survival of the remaining population (adults and birds in their first year) over the following winter, presumably because competition for winter food had been reduced. In this situation it seems that winter food availability (or possibly access to winter roost sites) is the factor which limits growth of this tit population. But populations may be limited by a range of factors. In one of the Scottish Blue Tit populations with which I am familiar, the population was not limited by food availability during the winter (I personally provided more than was required at feeding stations), but was limited by competition for nest sites—by putting out large numbers of nest boxes I was able to record a significant increase in the local population. You may recall from Box 5.7 that Jan Komdeur and colleagues were able to increase the world population of Seychelles Warbler by increasing the number of potential breeding territories for the birds. We will consider competition and population regulation again later in this chapter when we consider its impact upon bird community structure.

Flight path: space to breed and availability of nest sites and/or territories can limit productivity and drive changes in mating systems. Chapter 5.

Box 7.1   Aliens, pathogens, and competition

During the second half of the nineteenth century House Sparrows *Passer domesticus* were introduced to several sites across the USA. Today this successful generalist is one of the most common birds in the USA. It is a pest in many contexts and is held responsible for preventing the expansion of populations of native species because it outcompetes them for resources. However, in the second half of the twentieth century, things started to go badly for the sparrow; it lost ground to another alien invader—the House Finch *Carpodacus mexicanus*.

House Finches were translocated from their native western USA to Long Island in the east in 1940. This new eastern population struggled to get a foot-hold initially, but eventually it did establish itself and then quickly spread throughout the eastern states. As the finches spread it was noted that where they coexisted

sparrow numbers seemed to be falling. Was this more than a coincidence? The diets of the two species are very similar (both eat mainly seeds and vegetation); they are known to fight over resources when they meet; and there were reports that finches were breeding at sites traditionally occupied by sparrows. It certainly looked like the finches were competitively dominant with respect to the sparrows.

Evidence to support this hypothesis came as a result of an unusual natural experiment, one which in itself provides an excellent example of the impact of pathogens as a cause of density-dependent population decline. In the 1990s the eastern House Finch population was still growing strongly and expanding its range, but then in the winter of 1993–1994 birders in the state of Maryland began to report cases of a House Finch-specific conjunctivitis. Dubbed House

Finch disease, this was found to be caused by a bacterium, *Mycoplasma gallisepticum*, a pathogen previously restricted to domestic poultry. Subsequent research revealed that the disease was highly contagious and highly pathogenic—as Figure 7.2 shows, as many as 60% of the birds in infected populations died within 5 years of the disease emerging. It also spread very quickly, having reached Texas, Missouri, and Minesota by 1997.

It seems likely that this pathogen was able to spread so quickly precisely because of the ecological traits that made the finches such a successful species in the first place—their ability to tolerate one another and feed in large compact flocks (sites of infection) and their ability to disperse rapidly and colonize new areas (enhancing the geographical spread of the disease).

What about the sparrows? Remember that as the House Finch population increased a decrease in sparrow numbers was recorded. If the population dynamics of the two species are interlinked such that finch numbers are the check on the sparrow population we should expect to see the sparrows, being immune to *Mycoplasma gallisepticum*, to bounce back once the competitive pressure was released. This

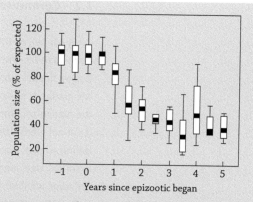

Figure 7.2 Changes in House Finch abundance following the emergence of *Mycoplasma gallisepticum* in a population. The data are expressed as changes relative to 100% (year -1, or 1 year prior to pathogen emergence). From Hochachka W.M. and Dhondt, A.A. (2000) Density-dependent decline of host abundance resulting from a new infectious disease. *Proceedings of the National Academy of Sciences, USA* **97**, 5303–5306.

is exactly what has been recorded by the veritable army of professional and amateur ornithologists in the USA who take part in the annual Christmas Bird Count (Figure 7.3).

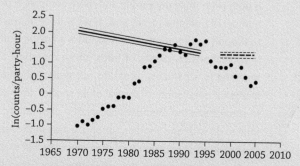

Figure 7.3 Abundance (log transformed) of House Sparrows (solid line and dotted line) and House Finches (points) recorded in the Christmas Bird Count (1970-2005). The shift from a solid to a dotted line in the case of the sparrows indicates the change in the direction of their population trend coincident with the emergence of *Mycoplasma gallisepticum* in 1993/4. From Cooper, C.B., Hochachka, W.M., and Dhondt, A.A. (2007) Contrasting natural experiments confirm competition between House Finches and House Sparrows. *Ecology* **88**, 864– 870.

7.2 Communities

In my garden, on the edge of a small village in the north of England, I might record as many as ten species of bird in a day and as many as 15 over the course of a year; the biggest flocks I record are of between six and ten House Sparrows *Passer domesticus*. My garden is not particularly rich in birds but it does provide the requirements of the particular community found there. It provides water (I have a small pond), food (I put out seeds, tolerate some weeds, and the plants that I cultivate support a bewildering array of insect pests), and places to breed (I put out nest boxes and birds nest in shrubs, trees, and under my eaves and roof tiles). Life for birds in my garden is not, however, without risk; my neighbour's cat no doubt takes her share of fledglings.

Just 3 km from home is my local nature reserve which, although small, supports very many more birds. This site, which is also on the edge of a village, consists of patches of reed- and rush-fringed open water, a small woodland, wet grassland, and muddy flashes. I regularly record more than 30 species in a visit and more than 60 species over the course of a year. Here I record flocks of 20 or 30 Tree Sparrows *Passer montanus* and of 50–100 starlings regularly. So why are these two bird communities so different? Essentially the nature reserve provides birds with the same resources as my garden—food, water, and nest sites—but of course it does it on a different scale. It provides a wider range of seeds and fruits, insects, small mammals, fishes, and amphibians as potential food. There are boxes, shrubs, trees, scrub, ditches, and banks for nests; islands for safe roosting by gulls; mud and wet grassland for feeding waders; shallow water and submerged vegetation for dabbling duck; and deep water for diving duck and grebes. The nature reserve is both larger than my garden and more complex ecologically. It provides a greater number of potential niches and, as we will see below, it therefore has the potential to support a community which includes a greater number of species.

Populations of species of birds (those members of a species living and interbreeding in a particular area) do not generally exist in isolation. They each live alongside populations of several other species, forming multispecies communities. In some cases the member species of a community may initially appear to have little to do with one another, simply existing in the same place. In others, however, the relationship between species is very clear. For example, in Scandinavian woodlands, Coal Tits *Periparus ater* and Pygmy Owls *Glaucidium passernium* are members of the same bird community, and one (the tit) is the food of the other. The owl has the same relationship with Willow Tits *Poecile montanus* in the community, and as we will see later in this chapter the ecologies of the two tit species interact such that the density of one has an effect upon the population of the other (but this is not a predator–prey relationship).

Typically a bird community will consist of a small number of species that are very numerous, and smaller numbers of less common species. Each of these will have a particular ecological role and will be an intrinsic part of the ecological network of which the community is a constituent part. As we have already seen, some species will be the food of others, some will prey upon non-avian animal members of the network's food web, others will be plant 'predators', but some will also be pollinators and seed dispersers.

Communities are dynamic

Although I have a pretty good idea that the community of birds in my garden will be the same tomorrow as it was today, it would be wrong to think of communities as being fixed. Just 10 years ago the colourful Goldfinches *Carduelis carduelis* that are now one of the most commonly recorded birds in my garden would have been an unusual visitor. Fifty years ago the Collared Doves *Streptopelia decaocto* that wake me in the morning would have been unknown in our village. These doves expanded their range northwards to colonize most of Europe over the course of the twentieth century. The goldfinches are thought to have moved into gardens relatively recently as a result of a change in human bird-feeding behaviour—a shift away from traditional peanut feeders in gardens towards a more varied range of bird foods. In the UK today niger seed and special feeders to dispense it are marketed as a highly effective goldfinch attractant. On the other hand, Song Thrush *Turdus philomelos* are now far less common in gardens in the UK generally, and in my area specifically, than they once were.

Communities change as a result of the immigration of new species and the extinction of old ones, and community stability is presumed to be achieved when immigration and extinction rates are balanced (see Figure 7.4).

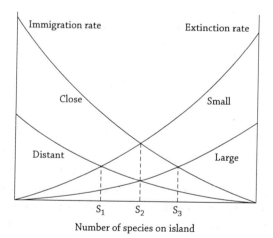

Number of species on island

Figure 7.4 The number of species on an island of a given size represents a balance between rates of immigration and extinction (s_1, s_2, s_3). The equilibrium point (number of species present) will vary in relation to both the size of the island and its proximity to a mainland. From Ricklefs, R.E. (1976) *The Economy of Nature*. W.H. Freeman, New York.

This apparent community stability is the basis of a concept termed equilibrium theory, one of a group of ecological theories developed as a result of studies that have been made of island systems and are often collectively referred to as island biogeography. Although I don't want to discuss the theories themselves in particular detail (interested readers should consult MacArthur and Wilson's *The Theory of Island Biogeography*), I do want to pick out a few key points. The first of these is illustrated in Figure 7.5A which shows that the bird communities of bigger islands typically have more species than do those of smaller islands. Figure 7.5B provides part of the explanation for this phenomenon. Ding and colleagues, who as would be expected also found that there were more species of birds on larger East Asian islands, have shown that species richness is also positively related to a measure of habitat heterogeneity, in this case an index of vegetation productivity. We will return to this link between species diversity and habitat diversity later in this chapter.

In the case of the data presented in Figure 7.5A and B, the islands concerned are real oceanic islands, i.e. land surrounded by sea, but if we were to think of islands as being one habitat surrounded by another we can shift our thinking to consider woodland fragments as islands in a 'sea' of grassland for example, and, as Figure 7.6 shows, the species–area relationship still applies. When Robyn Wethered and Michael Lawes surveyed the bird communities of fragmented montane forests in South Africa they found a very strong relationship between forest fragment area and the number of bird species that fragments supported.

The second point that I want to draw from the theory of islands' biogeography concerns island isolation. You will recall (from Figure 7.4) that islands closer to one another, or to a mainland (or, in the case of habitat islands, those that are part of a matrix of similar habitats or close to a significant contiguous area of that habitat), have a greater potential for colonization by new species. This means that in addition to island size, island connectivity is also a key determinant of observed species richness.

Whilst carrying out a study of the woodland bird communities of The Netherlands, van Dorp and Opdam have also confirmed that woodland patch size is the single most significant determinant of bird community size, but they have also demonstrated the importance of habitat connectivity and the proximity of patches to one another (Figure 7.7). In 1987, when they conducted their work, only 8% of The Netherlands were wooded and all of the woodland fragments were scattered throughout an agricultural landscape. However, in some parts of the country, the density of small woodland fragments was greater than in others, i.e. fragments were closer to one another, and were to some extent connected to one another by wooded ditches. Van Dorp and Opdam found that those woodlands that were part of a matrix of similar habitats interconnected by wooded ditches supported a richer bird community than did very isolated woodlands.

Key reference

MacArthur, R.H. and Wilson, E.O. (1967) *The Theory of Island Biogeography*. Princeton University Press, Princeton.

Flight path: adaptive radiations facilitate niche divergence and specialization. Chapter 1.

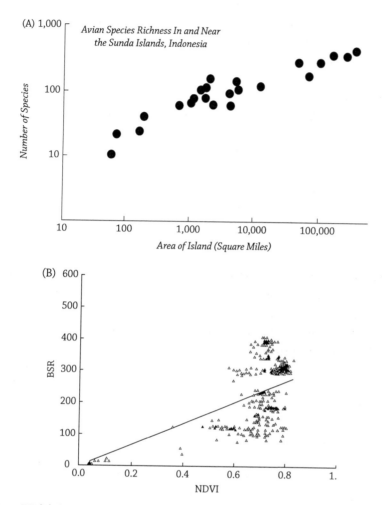

Figure 7.5 (A) The number of species of bird present on each of the 23 Indonesian islands listed is proportional to the area of that island. Smaller islands such as Christmas Island, which has an area of less than 100 square miles, support fewer species. As island area increases so does avian species richness, and New Guinea, the largest of the islands, supports the greatest number of species. Adapted from MacArthur, R.H. and Wilson, E.O. (1967) *The Theory of Island Biogeography*. Princeton University Press, Princeton. (B) The numbers of species of birds (BSR) for East Asian birds is strongly related to an index of vegetation productivity (NDVI) which is in turn related to habitat heterogeneity. From Ding, T.-S., Yuan, H.-W., Geng, S., Koh, C.-N., and Lee P.-F. Macro-scale bird species richness patterns of the East Asian mainland and islands: energy, area and isolation. *Journal of Biogeography* **33**, 683–693.

Niche divergence

You will recall that in Chapter 1 we discussed the mechanisms of natural selection and adaptive radiation by which individual species within communities evolve to coexist by specializing in their ecology in some way so as to avoid or at least

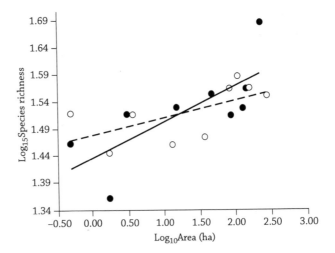

Figure 7.6 The number of species of bird present in a natural montane forest fragment increases with the area of the fragment. This relationship is apparent when the natural forest is surrounded by natural grassland (filled symbols, solid line) and when it is surrounded by a habitat matrix which includes artificial forest plantations (open symbols, solid line). From Wethered, R. and Lawes, M.J. (2003) Matrix effects on bird assemblages in fragmented Afromontane forests in South Africa. *Biological Conservation* **114**, 327–340.

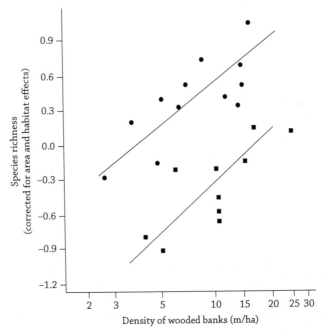

Figure 7.7 The density of wooded banks (a measure of habitat connectivity) is a key determinant of species richness in both the eastern Netherlands (top line) and the central/southern Netherlands (bottom line). From van Dorp, D. and Opdam P.F.M. (1987) Effects of patch size, isolation and regional abundance on forest bird communities. *Landscape Ecology* **1**, 59–73.

minimize interspecific competition. We saw, for example, that through beak specialization the Charidriiform waders are able to feed alongside one another in multispecies flocks on estuarine mud flats, and that the Geospzid finches of the Galapagos islands have evolved beak morphologies which allow them to coexist and exploit the full range of feeding opportunities available to them. This is possible because faced with an array of different feeding opportunities at a single site the birds are able to specialize, and each therefore occupies a different feeding niche. The kind of adaptive radiation exemplified by the Galapagos finches is possible because the ancestral finch colonized an island system that contained an array of vacant niches, and this is a fundamental part of the explanation for the species–area relationship previously discussed—quite simply larger island have more potential niches and so can support a greater number of species (see Figure 7.5).

Box 7.2 Niche segregation in a riparian community

The fast-flowing mountain streams of the Himalayas are thought to be home to more species of specialist riparian birds than any other river system. However, until Sebastian Buckton and Steve Ormerod made detailed observations of the behaviour of these birds and compared their body measurements, no ecologist had attempted to explain the mechanisms by which they coexist. Working in four valleys in central Nepal the researchers studied five species of insectivore: the Spotted Forktail *Enicurus maculates*, Little Forktail *Enicurus scouleri*, White-capped Water Redstart *Chiamorrornis leucocephalus*, Plumbeous Water Redstart, *Rhyacornis fuliginosus*, and Brown Dipper *Cinclus pallasii*. They captured and measured examples of each species, collected their faeces (to determine diet), and recorded how each of them used riverine microhabitats whilst foraging.

They found that the birds spent between 50% (Brown Dipper) and more than 80% (forktails) of their time foraging, but that the species foraged in different ways and in different places from one another, presumably therefore minimizing competition. The Brown Dipper was the only species to forage underwater (a behaviour characteristic of dippers the world over), and was observed at both the edges of streams and in their centres. Similarly,

Plumbeous redstarts foraged along stream edges and in stream centres where they tended to fly-catch above dry boulders and occasionally pick between them. In contrast, White-capped Redstarts favoured stream margins where they foraged amongst dry boulders, only occasionally fly-catching or picking food from the splash zone. The two forktails also exhibited microhabitat-based segregation; the Spotted Forktails preferring dry areas and litter, and the Little Forktails utilizing the splash zone where they fed on wet boulders and submerged gravels. Faecal analysis revealed that by using the available microhabitats in this way, the five species were able to some degree to avoid competition and specialize to some extent in their diets (see Figure 7.8). Although it is clear from the figure that there is considerable overlap in prey taken, it is also clear that the two more aquatic species, Brown Dipper and Plumbeous Water Redstart, show least overlap and the three less aquatic species are also separated from one another. Furthermore, the sizes of individual prey items selected was also found to contribute further to dietary separation. For example, although both Brown Dipper and Little Forktails fed on *Ephemeroptera*, those individuals eaten by dippers were considerably larger.

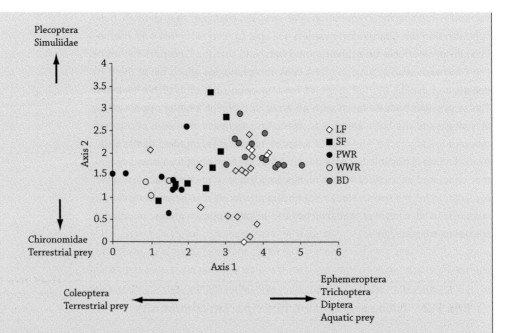

Figure 7.8 Prey selection as an indicator of niche segregation in Little Forktail (LF), Spotted Forktail (SF), Plumbeous Water Redstart (PWR), White-capped Water Redstart (WWR), and Brown Dipper (BD). From Buckton, S.T. and Ormerod, S.J. (2008) Niche segregation of Himalayan river birds. *Journal of Field Ornithology* **79**, 176–185.

It is clear, therefore, that the members of this particular community of birds are able to coexist because they each specialize to some extent in aspects of their foraging behaviour. In this particular example the researchers did not observe interspecific aggression and so it seems likely that the species have evolved to complement one another rather than having achieved resource partitioning as a result of intense competition.

Although we often focus on feeding behaviour when we discuss the niche, it is in fact a far broader concept. Each species of bird has a fundamental niche which can be described as being the ecological space in which it can exist (where it is, what it does, what it eats, etc.). However, when we consider the ecology of birds as members of particular ecological networks and avian communities we most commonly record a sub-set of the fundamental niche termed the realized niche. This is that part of the fundamental niche that the bird can actually use at that time—being effectively squeezed as it is between other competing species.

Niche shifts, ecological release, and competition

The competitive exclusion principle dictates that no two species may occupy the same niche at the same time and in the same place, and so where species overlap in range they have evolved competitively or through complementarity to each occupy a smaller realized niche. This is perhaps most apparent when a community shift occurs and competitive pressure is removed, with the result that a species expands its ecology and moves beyond the confines of the realized niche. For example, where they co-occur. Yellow-rumped Warblers *Dendroica coronata* and Black-throated Green Warblers *Dendroica virens* exhibit niche segregation and both utilize different areas of the spruce trees in which they forage. The Yellow-rumped Warblers concentrate their activity in the bottom 6 feet or so of the tree, whilst the Black-throated Green Warblers feed in the top part of the tree. In fact there can be other species of *Dendroica* warbler using the same trees, and under such conditions foraging area specialisms can become even more marked. In some parts of their ranges these two species do not co-occur, and when the Black-Throated Green Warbler is absent the Yellow-rumped Warbler undergoes a niche expansion (sometimes referred to as a competitive release) and forages 30% higher into the tree. Interestingly when the situation is reversed and it is the Yellow-rumped Warbler that is absent, the Black-throated Green warbler does not exhibit a shift in niche. This suggests strongly that this is behaviourally and ecologically the dominant species and that its presence dictates the realized niche breadth of the Yellow-rumped Warbler.

The fact that one member of the spruce warbler community dictates the feeding opportunities of another is an excellent example of the role of competition in structuring bird communities. It is often the case, however, that community structure is also strongly influenced by a range of factors that have a cumulative effect. For example, Cecilia Kullberg and Jan Ekman have shown that the structure of the European tit community found on Scandinavian islands is a result of an interaction of two forms of interspecific competition (exploitation competition and interference competition) and the actions of a predator. The communities that they have studied include four species: the diminutive Coal Tit *Periparus ater*, the larger Crested Tit *Lophophanes cristatus*, the Willow Tit *Poecile montanus*, and the Pygmy Owl *Glaucidium passerinum* which is a tit predator (Plate 28).

Throughout Scandinavia these tits are found sympatrically, and when they are the birds segregate when feeding in a manner similar to the *Dendroica* warblers that we have just discussed. The smaller Coal Tits forage closer to branch tips and the Larger Crested and Willow Tits utilize the areas of the tree closer to the trunk. But on islands where there are only Coal Tits present they forage throughout the tree, having undergone an apparent niche expansion.

Coal Tit-only islands are occasionally colonized by either Willow or Crested Tits, but they seem to be unable to establish stable populations. Coal Tits are also more efficient foragers than either Willow Tits or Crested Tits and are superior in exploitation competition for food. The may also outcompete the larger species because they are more productive, having up to two large broods per season when Crested and Willow Tits manage only a single smaller brood each year.

Flight path: birds segregate when feeding to minimize competition. Chapters 1 and 6.

Key reference

Morse, D.H. (1980) Foraging and coexistence of spruce-wood warblers. *Living bird* **18**, 7–25.

Key reference

Kullberg, C. and Ekman, J. (2000) Does predation maintain tit community diversity? *Oikos* **89**, 41–45.

Flight path: the presence of a predator can alter feeding behaviour. Chapter 6.

However, the situation changes when Pygmy Owls are present on an island. In these situations, the Larger Crested and Willow Tits exploit their social dominance over Coal Tits and monopolize the safer feeding areas close to the tree trunk through interference competition. This forces the Coal Tits to forage in risky areas on the outside of the tree, and as a result they suffer disproportionately from owl predation. Because the owls limit the Coal Tit population exploitation, competition becomes irrelevant and a stable community of the three tit species becomes possible. In this situation, the Pygmy Owl is clearly acting as a 'keystone' predator, and its actions maintain community diversity.

7.3 Extinction and conservation

When a population shrinks below a minimum viable size it becomes effectively extinct. It no longer fulfils an ecological role and it no longer has the capacity to recover and grow. Without human assistance (and in many cases even with it) such populations shrink until there are no birds left and they become extinct in the traditional sense of the word. If another population of the same species exists in another place, there is the chance that natural re-colonization will occur, as has happened in Scotland in the case of the osprey. This species became extinct in Scotland in 1916 (having already been lost in England in 1840), but in 1954 a pair returned to breed. Since that time, thanks to the conservation efforts of many individuals and organizations, the British population has grown to around 150 pairs.

If natural re-colonization is unlikely, then conservation efforts might result in the successful re-introduction of the species, moving members of a surviving population into the area to be colonized. This strategy was used in an attempt to re-introduce the osprey to southern England when it became apparent that a rapid natural spread southwards from Scotland was unlikely.

However, as we all know, there comes a point when a species goes the way of the Dodo *Raphus cucullatus*, the Passenger Pigeon *Ectopistes migratorius*, and the Great Auk *Penguinius impennis*, when the last population shrinks to the point that the last individuals die, then the species becomes extinct and is extinct for ever. In 2006 the IUCN (International Union for the Conservation of Nature) stated that 135 bird species had become extinct since the year 1500. A sobering thought, particularly when one takes into account the rate of increase in extinction rate over the same time period (Figure 7.9).

Clearly small populations are particularly vulnerable to extinction, but small numbers alone, however, do not explain what makes a species vulnerable to extinction. After all, the Passenger Pigeon went from being one of the most numerous birds known to being extinct in a very short period—birds that are hunted by man, but which man makes no effort to harvest in a sustainable way are at particular risk.

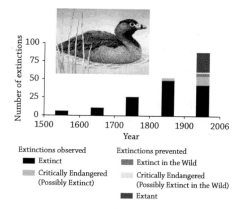

Figure 7.9 Estimated numbers of bird extinctions and of numbers of critically endangered bird species in the past five centuries. The inset bird, the Atitlàn Grebe *Podiymbus gigas*, was found on lake Atitlàn, Guatemala. It became extinct in 1986. From Rodrigues, A.S. (2006) Are global conservation efforts successful? *Science* **313**, 1051–1052.

A number of factors make one species more likely to become extinct than another, and when a number of these coincide the odds are not good.

For example, species with a very narrow geographical or ecological range are at particular risk, birds found on just one oceanic island, or extreme specialists for example. Loss of habitat in such a situation could be catastrophic. Conservation efforts can, however, make a difference in such situations. Recall that Jan Komdeur and his colleagues greatly increased the world population of Seychelles Warbler by translocating a group of birds to a second island (see Chapter 5).

However, even though the threat to the warblers is lessened, these small populations remain particularly vulnerable. There is often little scope for movement of individuals between such populations, and each is in itself vulnerable by virtue of its small size. For example, Jared Diamond and coworkers have reported that of the birds in the Bogor Bogor Arboretum and Botanical Garden, Java, a green space which had been isolated from similar habitats for 50 years at the time of Diamond's work, 75% of those species with a small initial population became extinct, whereas all of the species with an initially large population survived.

Small populations, or populations that have rapidly grown from a small founder population, face another problem—reduced genetic diversity—which may make them more susceptible to emergent diseases because they are unable to evolve resistance (this could, for example, explain the susceptibility of the House Finch to *Mycoplasma gallisepticum*).

Species with low population densities, or large range requirements (such as the larger birds of prey), are particularly vulnerable; their habitats are susceptible to fragmentation, and encounters with potential mates can become infrequent. We have already seen that a species with a naturally low rate of productivity is less able

Key reference

Diamond, J.M., Bishop, K.D., and van Balen, S. (1987) Bird survival in an isolated Javan woodland: island or mirror? *Conservation Biology* **7**, 39–52.

to expand its population; similarly, species with low dispersal potential are at risk. They simply lack the ability to spread. Island rail species are an excellent example of this phenomenon. Many have lost the ability to fly (in some cases they are psychologically flightless even though they retain the mechanical ability). Most, if not all, are threatened with extinction.

Migrants rely upon multiple locations and stopping off points between them, and so are particularly susceptible to habitat loss. They are often funnelled through geographical bottlenecks where predators can concentrate their effect and, in some cases, such as the North American Warblers, have very restricted wintering grounds. A single hurricane could make a species extinct just because it destroys the forests of a single Caribbean island. This scenario is not at all far fetched—Bachman's Warbler *Vermivora bachmanii* is thought to have become extinct in the 1960s as a result of the deforestation of Cuban forests to make way for sugar cane plantations.

Conservation can be a success

Earlier in this chapter I cited the sobering statistics provided by Ana Rodrigues which bring home the scale of the threat of extinction faced by birds. However, we should not lose hope. Throughout this chapter and elsewhere in this book I have described examples of monitoring programmes, nest box schemes, habitat management, breeding programmes, re-introductions and translocations, and other practical conservation efforts. The human love of, and admiration for, birds is such that armies of citizen scientists can be mobilized across the globe to raise funds

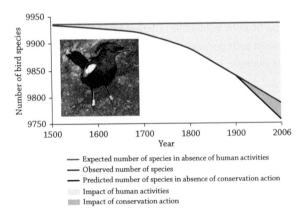

Figure 7.10 The estimated impact of global conservation actions to prevent bird extinctions. In the last 100 or so years, conservation efforts have resulted in more than 30 species being brought back from the brink of extinction. The Seychelles Magpie-Robin (inset) was saved from extinction by an integrated programme of translocations, habitat management, and predator eradication. From Rodrigues, A.S. (2006) Are global conservation efforts successful? *Science* **313**, 1051–1052.

and carry out practical work in an effort to conserve birds and thereby conserve the ecosystems that both we and birds are part of (Plate 29). But is it working? I think that it is. Ana Rodrigues provides evidence that the tide is turning (Figure 7.10), and at the scale of my own experience I can see birds in the UK today that were locally extinct in my memory and that conservation efforts have restored to us. As a bird ringer and birder I can see the passion and dedication of private individuals for their own conservation actions, and as an academic I am aware of the advances in our understanding that leaders in the field of ornithology are making.

It is now almost 50 years since Rachel Carson brought to the attention of the world the threat of a 'silent spring'. We may not have solved the problem yet—but working together we are inching closer to doing so.

> **Key reference**
>
> Carson, R. (1962) *Silent Spring*. Houghton Mifflin, Boston.

Summary

The life history strategies of individual species influence their population size, as do numerous environmental pressures both natural and anthropogenic. Communities of birds are able to coexist because the species within them exhibit niche separation and competition avoidance. Many of the world's bird are threatened as a result primarily of human activity, but by working together species can be saved from the brink of extinction.

Questions for discussion

1. What factors limit bird populations?
2. What steps should be taken to conserve bird populations and bird species?
3. Are extinctions inevitable?

Index

Page numbers in **bold** refer to information in boxes or illustrations